ABOVE GROUND STORAGE TANKS

Practical Guide to Construction, Inspection, and Testing

ABOVE GROUND STORAGE TANKS

Practical Guide to Construction, Inspection, and Testing

SUNIL PULLARCOT

CRC Press is an imprint of the
Taylor & Francis Group, an **informa** business

CRC Press
Taylor & Francis Group
6000 Broken Sound Parkway NW, Suite 300
Boca Raton, FL 33487-2742

First issued in paperback 2020

ISBN-13: 978-1-4822-2202-9 (hbk)
ISBN-13: 978-0-367-73829-7 (pbk)

Visit the Taylor & Francis Web site at
http://www.taylorandfrancis.com

and the CRC Press Web site at
http://www.crcpress.com

Contents

Preface

A majority of above ground storage tanks (especially those used for storage of hydrocarbon liquids) are constructed according to API (American Petroleum Institute) standards. Apart from these standards, there are many books available on the market that deal with the engineering aspects of storage tanks and also with their layout and safety-related matters. However, there no book has been published that deals with all aspects subsequent to design, which basically covers all site construction activities through commissioning, including inspection and testing. This inspired the author, Sunil Pullarcot, to compile this book, *Above Ground Storage Tanks: Practical Guide to Construction, Inspection, and Testing*.

Because API Standard 650 (*Welded Tanks for Oil Storage*) is the predominant standard that is followed in the construction of storage tanks, this book was written based on the requirements of the API 650 March 2013 edition, which is the current edition.

This book deals with every aspect that a construction team needs to know after receipt of an order to construct a storage tank at site. It describes all the activities that are to be carried out in a sequential manner. Although this book also touches on the design of foundations and mechanical design of storage tanks, these two topics are given only a superficial treatment and to provide an overview according to a preferred work sequence at the site.

The design of the civil foundation is usually carried out by a civil engineering team, and the design of the tank is done by a mechanical group. This book even gives an overview of these two aspects while sequentially progressing from design through construction up through commissioning activities.

This book principally targets beginners in the construction industry, and deals with practical tips to construct a tank according to dimensional tolerances and quality as required by API 650. Many of the tips included herein are from the author's practical improvisations of existing practices within the industry that were proven successful. In this regard, the author fondly remembers the immense support he received from renowned contractors such as Petrofac and the various inspection agencies and personnel he worked with, giving this book validity by way of their valuable opinions and advice.

Above Ground Storage Tanks deals with the mechanical construction, inspection, and testing of storage tanks and the stages after design, meaning that design and preparation of drawings are not covered. API 650 stipulates an array of requirements to be taken care of during design and construction and accommodates further requirements from clients as well (especially those in the oil and gas industry). However, these standards and specifications are practically silent on how to achieve these requirements during the process of transforming drawings into real structures. This book provides this vital information, but the author was careful not to overspecify requirements, because this unnecessarily adds to the costs. However, the author was selectively cautious and stringent in specifying requirements with the intention of avoiding reworks and repairs, which eventually work out to be more expensive.

As indicated, the main highlight of this book is the satisfactory explanation the author tries to provide to all construction-related aspects of storage tank erection by explaining the logistics and rationale behind the requirements spelled out in applicable codes and specifications. Apart from this, the book also contains various summaries of requirements as a quick reference to code and client requirements.

This book also provides logical explanations of various code requirements and is capable of throwing some light on the unexplained side of storage tank codes. This feature of the book makes it unique when compared with other books commonly in use.

On the basis of the author's abundant practical experience, as in the case of his first book, *Practical Guide to Pressure Vessel Manufacture* (Marcel Dekker, Inc., 2002), this book can also be considered an effort to bridge the gap between standards and specifications and the actual construction of storage tanks taking place at the site.

1

Storage Tanks

1.1 Introduction

A *storage tank* is a container, usually for holding liquids and sometimes compressed gases (gas tank). The term can be used for reservoirs (artificial lakes and ponds) and for manufactured containers. The usage of the word *tank* for reservoirs is common or universal in common parlance.

Storage tanks operate under no (or very little) internal pressure, distinguishing them from *pressure vessels*. Storage tanks are often cylindrical in shape, are perpendicular to the ground with flat bottoms, and have a fixed or floating roof. There are usually many environmental regulations applied to the design and operation of storage tanks, often depending on the nature of the fluid contained within. Above ground storage tanks (ASTs) differ from underground storage tanks (USTs) with regard to design considerations and thereby applicable regulations as well.

As mentioned, tanks are meant to carry large quantities of liquid, vapor, or even solids for a variety of process applications. The process applications include the following, apart from pure storage function:

1. Settling
2. Mixing
3. Crystallization
4. Phase separation
5. Heat exchanging
6. Reactors

However, storage in large quantities is the principal purpose of tanks. Storage may be for a sales network wherein there shall be enough stock to cater to demand without any break, or storage may be for a process plant to facilitate uninterrupted working of a downstream plant or intermediate storage within the plant to store intermediate products for a short duration of time. The above-mentioned parameters are of interest to the designer in arriving at a rational size for a tank and in designing sizes of the nozzle, the piping in and out, and other safety attachments. The coverage of this book is mainly related to construction activities of storage tanks at site, upon completion of design of the same, including the mechanical design, implying that the process (sizing) and mechanical design of the tank are frozen and drawings have Approved for Construction (AFC) status. Therefore, this book covers all activities in construction of storage tanks right from the laying of bottom plates, through shell and roof erection, up to surface preparation and coating of the completed tank prior to handing over for commissioning.

Whatever use tanks are subjected to, the construction methodology that is followed is more or less the same, and differences if any in methodology shall be because of the configuration of the tank, which is marginal in nature.

As the present-day need for storages is continuously increasing because of increasing consumption, tanks of larger and larger capacities are built. These are the challenges of the time, and technocrats had risen to the expectation and were able to formulate codes, standards, and regulations for construction of such large storage tanks. The requirements spelled out in such codes and standards are based on many years of research, supplemented by hands-on experience of experts in the field. Because of this reason, construction methodology has changed a lot over the years. Similarly in the design side also, tremendous changes have taken place over time, due to research about materials and their properties, which are implemented through periodic revision issues to codes and standards. Furthermore, as approximations give way to certainties, Safety factors are being reduced continually in the design of storages. In other words, as time passes, engineering is moving from a state of ignorance to an era of knowledge and enlightenment.

While talking about construction of storage tanks, it should not be forgetten that their design needs to be carried out as a prelude to this activity. Since the requirements vary from place to place and process to process, based on economic considerations, every tank (except identical tanks in the same vicinity) requires a separate design, which makes it unique and tailor-made for the particular use.

As is known, tanks can be constructed only on a very strong foundation; failing this, the stability of the tank can be adversely affected, ending up in a disaster due to the large volume of liquid handled, which exerts tremendous forces on the foundation beneath the tank. While the design of the foundation and its construction fall under the realm of civil engineering, all subsequent design and construction activities (mechanical design and construction) of the storage tank are undertaken by the mechanical group. Because of this jurisdictional demarcation, after all civil works and associated quality control checks on the foundation are completed, it has to be handed over to the mechanical team. Apart from the tests carried out on the foundation, the first comprehensive evidence of its adequacy is obtained from foundation settlement measurements carried out during hydrostatic testing of the storage tank. While uniform settlement is tolerated, uneven settlement needs further investigation and evaluation to assess the safety and integrity of the constructed storage tank.

1.2 Classification of Storage Tanks

One of the fundamental classifications employed in tankages is based on the location of the tank whether it is above ground or below, whose design and manufacturing features vary drastically. The number of above ground tanks is far in excess of those below ground, and they have their bottom directly resting on an earthen or concrete foundation. In some specific cases, tanks were placed on grillages, which facilitates periodic inspection of the bottom plates of the tank.

When the question of providing a storage facility arises, the obvious choice will be above ground storage because of the following reasons:

1. Above ground storage tanks are easy to construct.
2. The cost associated is low compared to that of other modes of storage.
3. Larger capacities are possible.

Another classification is elevated tanks placed on structural supports. Water supply tanks placed on steel structures belong to this category.

Underground storage tanks are usually designed for capacities in the range of 23 m^3 to 91 m^3 (5,000 gallons–20,000 gallons), with the majority falling under 55 m^3 (12,000 gallons). These are mainly used to store fuels and a variety of chemicals.

Since this book deals with construction aspects related to above ground storage tanks, the following chart shows a broad classification of storage tanks, with detailed classifications of above ground storage tanks.

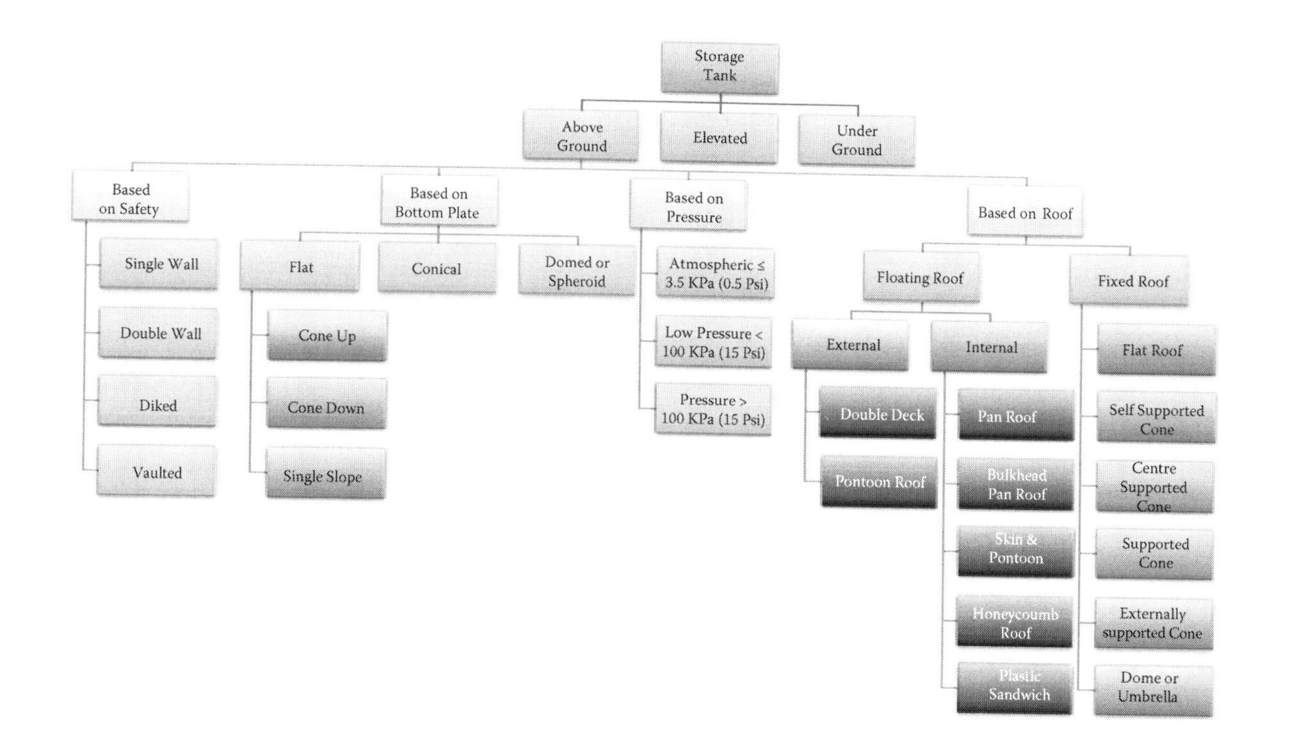
Storage Tank
Above Ground
Elevated
Under Ground
Based on Safety
Single Wall
Double Wall
Diked
Vaulted
Based on Bottom Plate
Flat
Cone Up
Cone Down
Single Slope
Conical
Domed or Spheroid
Based on Pressure
Atmospheric ≤ 3.5 KPa (0.5 Psi)
Low Pressure < 100 KPa (15 Psi)
Pressure > 100 KPa (15 Psi)
Based on Roof
Floating Roof
External
Double Deck
Pontoon Roof
Internal
Pan Roof
Bulkhead Pan Roof
Skin & Pontoon
Honeycoumb Roof
Plastic Sandwich
Fixed Roof
Flat Roof
Self Supported Cone
Centre Supported Cone
Supported Cone
Externally supported Cone
Dome or Umbrella

1.3 Classification of Above Ground Storage Tanks

Prevailing codes, standards, and regulations basically classify storage tanks according to their *internal pressure*, as shown in the following table.

Classification	Description
Atmospheric pressure tanks	Tanks operating at an internal pressure slightly above atmospheric, up to a pressure of 3.5 kPa (0.5 psig).
Low-pressure tanks	Tanks operating at a pressure higher than that of atmospheric storage tanks, up to a pressure of 100 kPa (15 psig).
Pressure tanks	Tanks operating at a pressure above 100 kPa (15 psig), normally called *pressure vessels*. The term *high-pressure tank* is not used by those working with these tanks, as its falls under the specialized category of pressure vessels. The considerations for pressure vessels are totally different, as are the applicable codes, standards, and regulations.

Another way of classification is based on the *type of construction* of the storage tank.

Classification	Subclassification	Further Classification	Description
Fixed roof tanks	Flat roof		The flat roof is for small-diameter storage tanks without any supports.
	Self-supported cone roof		The roof is conical in shape but is self-supporting due to the rigidity offered by the shape of the roof.
	Center-supported cone roof		The cone roof is supported at the center and is usually used for medium-sized tanks.
	Supported cone roof		The roof is supported at many points as needed to take care of the load of the roof as well as other mobile and static loads expected.
	Externally supported cone roof		The roof of this type is externally supported even by means of external columns outside the tank shell.
	Dome or umbrella roof		The peculiar shape makes the roof a bit more rigid, but it is always expensive compared to other types of roofs.

(*Continued*)

Classification	Subclassification	Further Classification	Description
Floating roof (FR) tanks	External FR	Pontoon roof	This is common for a floating roof for sizes in the range of 10 m to 30 m (30 ft–100 ft) in diameter. The roof is simply a steel deck with an annular compartment that provides buoyancy.
		Double deck roof	Double deck roofs are built for a wide range of diameters including those above 30 m (100 ft) in diameter. They are very strong and durable because of the double deck and hence are suitable for large diameter tanks.
	Internal FR	Pan roof	Pan roofs are made from a simple sheet of steel disks with the edge turned up for buoyancy. These roofs are prone to capsizing and sinking because a small leak can cause them to sink.
		Bulkhead pan roof	A bulkhead pan roof has an open annular compartment at the periphery to prevent the roof from sinking in the event of a leak.
		Skin and pontoon roof	Skin and pontoon roofs are usually constructed of an aluminum skin supported on a series of tubular aluminum pontoons. These tanks have a vapor space between the deck and the liquid surface.
		Honeycomb roof	A honeycomb roof is made from a hexagonal cell pattern that is similar to a beehive in appearance. The honeycomb is glued to top and bottom aluminum skins that seal it. This roof directly rests on the liquid.
		Plastic sandwich roof	A plastic sandwich roof is made from rigid polyurethane foam panels sandwiched inside a plastic coating.

Yet another classification of tanks is based on the *bottom plate* provided for the tank.

Classification	Subclassification	Description
Flat bottom	Flat	For tanks up to 6 m to 10 m (20 ft to 30 ft) in diameter; inclusion of a small slope does not provide benefit and hence a flat bottom is permitted.
	Cone up	The cone up has a high point at the center and a slope of about 25 mm to 50 mm per 3 m (1″ to 2″ per 10 ft) diameter.
	Cone down	The cone down slopes toward the center. A collection sump is usually provided at the center with piping under the tank to a well for draining purposes.
	Single slope	The single slope bottom is tilted to only one side. Drainage is taken from the low point. As the diameter increases, the difference in elevation of the bottom plate increases and hence the use of the same is limited to a diameter up to 30 m (100 ft).
Conical bottom	—	Conical bottom tanks provide complete drainage of even residues. Since these types are costly, they are limited to small sizes and are often found in the chemical industry or in processing plants.
Domed or spheroid	—	The domed or spheroid bottom also provides complete and effective drainage. It is costly and difficult to construct, so its use is restricted to smaller sizes.

Another type of classification for *small tanks* is based on the *safety feature* of the tank, as shown in the following table.

Classification	Description
Single wall tank	Usually cylindrical and either vertical or horizontal.
Double wall tank	Common for both above ground and underground applications, since the tank can contain leaks from the inner tank, and it serves as leak detection.
Diked or unitized secondary containment tank	The primary containment above ground storage tank is housed within a rectangular steel dike that contains the product spill in the event of a leak or rupture of the primary tank. The dike can be an open or closed type.
Vaulted tank	Refers to tanks installed inside concrete vaults.

1.4 Selection of Type

The decision to select a particular type of tank is principally based on process requirements. However, the cost of the tank also plays a very vital role in the selection process. For example, for storage of fire water, only fixed roof storage tanks are selected. Whereas in the case of storage for highly volatile fluids, floating roof tanks are preferred, in spite of the comparatively high cost of construction for floating roof tanks.

As mentioned earlier, design and manufacture of such storage tanks need special consideration, as follows:

1. Earth load is to be considered in the design of the tank and its foundation.
2. Buoyancy forces also shall be considered, especially in the case of buried tanks.
3. The requirements for anchoring also need to be considered in most of the buried tanks.
4. As the tanks are prone to external corrosion, the following matters shall be considered seriously:
   a. Backfill material and its properties
   b. Necessity of the cathodic protection system.
   c. Necessity for coating or lining of exterior surface
5. Necessity of leakage monitors as per regulatory requirements shall be considered.

1.5 Two Common Types of Storage Tanks

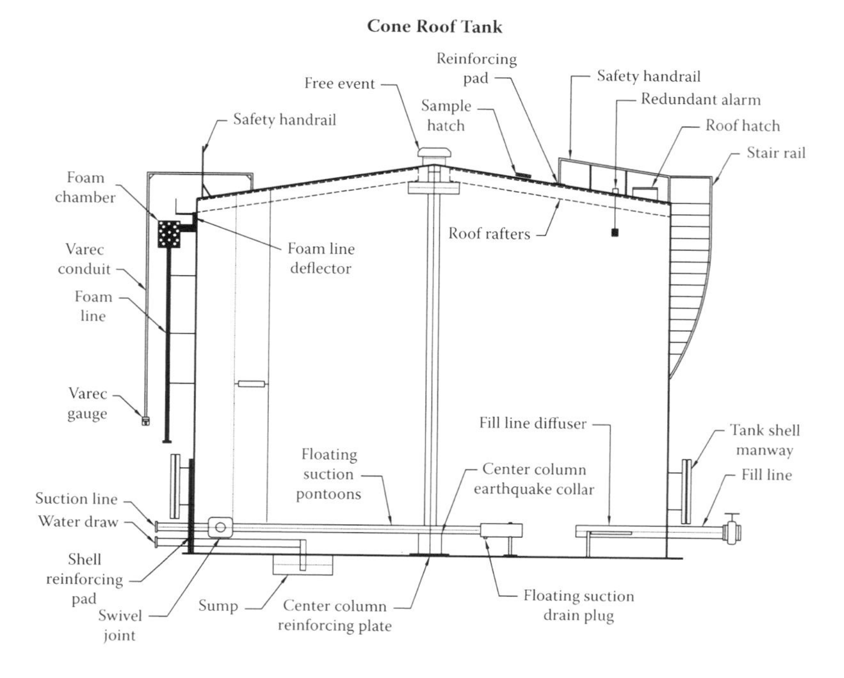

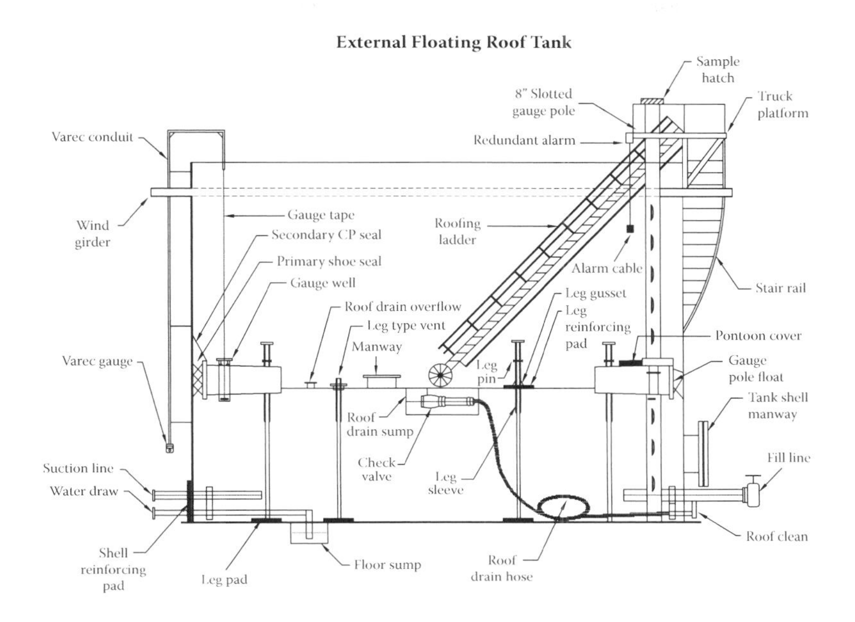

1.6 Design and Construction Standards for Above Ground Storage Tanks

As mentioned earlier, a storage tank is a container for storage of liquids or gases in large volumes. A tank may be constructed of ferrous or nonferrous metals, alloys, reinforced concrete, wood, and so on, depending on the use for which it is to be built. Though there are many codes prevailing for the design, manufacture, inspection, and testing of storage tanks, the most widely used across the world is that by the American Petroleum Institute (API).

1.7 API Specification for Storage Tanks

Storage tanks are part and parcel in the production, refining, transportation, and marketing of petroleum products for their uninterrupted operation. API maintains several documents that address the design, fabrication, operation, inspection, and maintenance of above ground storage tanks (ASTs) and underground storage tanks (USTs).

As in the case of other standards, API storage tank standards are also developed by committees consisting of experienced tank designers, fabricators, owners, and operators that bring a wealth of accumulated knowledge from their respective areas of expertise. Although API standards cover many aspects of AST and UST design and operation, they are not all-inclusive. Apart from API, there are several other organizations that also publish standards on tank design, fabrication, installation, inspection, and repair that may be more appropriate in some instances than API standards. Wherever API specifications are silent, related specifications and published materials are often referred to for more clarity on requirements, a few of which are provided below.

- American Society of Mechanical Engineers (ASME): www.asme.org
- American Society for Testing and Materials (ASTM): www.astm.org
- American Water Works Association (AWWA): www.awwa.org
- Building Officials and Code Administrators International (BOCA): www.bocai.org
- NACE International (Corrosion Engineers): www.nace.org
- National Fire Protection Association (NFPA): www.nfpa.org
- Petroleum Equipment Institute (PEI): www.pei.org
- Steel Tank Institute (STI): www.steeltank.com
- Underwriters Laboratories (UL): www.ulonet.ul.com
- International Fire Code Institute (Uniform Fire Code): www.ifci.com

As mentioned earlier, storage tanks are found in most sectors of the petroleum industry. These include exploration and production, refining, marketing, and pipelines. Storage tanks can be divided into two basic types: AST and UST. While ASTs are used in production, refining, marketing, and pipeline operations, USTs are typically used in marketing gasoline at retail service stations.

Though API has published standards for construction of ASTs, API did not publish standards for construction of USTs. The common standards used for the design and construction of these tanks are Underwriters Laboratories UL58, 1316, or 1746, and the Steel Tank Institute's F841-91.

For ASTs, the following standards were developed by API:

1. API Spec. 12B, *Bolted Tanks for Storage of Production Liquids*
2. API Spec. 12D, *Field Welded Tanks for Storage of Production Liquids*
3. API Spec. 12F, *Shop Welded Tanks for Storage of Production Liquids*
4. API Spec. 12P, *Fiberglass Reinforced Plastic Tanks*

5. API 620, *Design and Construction of Large, Welded, Low-Pressure Storage Tanks*
6. API 650, *Welded Steel Tanks for Oil Storage*

Tanks manufactured according to API 12 are much smaller than API 650 tanks and are often subject to different operating conditions.

Among the above-mentioned API standards, the predominantly used standard is API 650, *Welded Steel Tanks for Oil Storage*.

The principal aim of any standards is to provide the industry with tanks of adequate safety and reasonable economy for use for the service life of the process plant. Rather than having a standard for each type of industry, these standards have evolved for the use of many industries by compiling the basic essential requirements for a variety of industries such as fertilizer, petrochemical, and so on. Therefore, the standards provide a broad availability of proven, sound engineering and operating practices and hence do not obviate the need for applying sound engineering judgment regarding when and where the requirements are to be applied. As these standards provide only basic minimum requirements, specific requirements (if any are applicable to specific use) shall be explicitly indicated on a data sheet formulated by the owner or consultants while floating inquiries for storage tanks.

Furthermore, the standards thus evolved do not apply any restraint on the purchaser's prerogative to decide on size and capacity, based on process, safety, and economic considerations.

1.8 Jurisdiction of Various Standards for Storages

1.8.1 API 650: *Welded Steel Tanks for Oil Storage*

This standard provides minimum requirements for material, design, fabrication, erection, and testing for vertical, cylindrical, above ground, closed and open top, and welded storage tanks in various sizes and capacities for internal pressures approximating atmospheric pressure (internal pressures not exceeding the weight of the roof plates). Its Annex F specifies requirements for tanks with mild positive internal pressures.

Since the purpose of this standard is to construct storage tanks of adequate safety and reasonable economy for petroleum product storage, this standard does not provide any specific sizes for the tanks, which falls in the domain of end users, depending on their specific use.

1.8.2 API 620: *Design and Construction of Large, Welded, Low-Pressure Storage Tanks*

This standard covers large, field-assembled storage tanks intended to contain petroleum and petrochemical intermediates or products, such as LNG, LPG, ammonia, and so on in the liquefied stage. Since the internal pressure has limitations on account of large diameters, these gases (at normal atmospheric temperatures) need to be refrigerated to subzero temperatures to store as liquid under low positive pressures. This poses challenges with regard to suitable materials for such low temperatures and suitable welding and construction methodologies. Though such low temperature storage tanks are usually provided with a refrigeration system (in addition to insulation), the scope of API 620 covers only the tank proper. Here also, the purpose of the standard is to provide safe and cost-effective storage, and hence the standard does not specify the sizes. The mandatory sections of the standard specify requirements for storage at temperatures as low as –46°C (–50°F). Its Annex S covers stainless steel low-pressure storage tanks in ambient temperature service in all areas, without limit on low temperatures. Similarly, Annex R covers low-pressure storage tanks for refrigerated products at temperatures in the range of +4°C to –51°C (+40°F to –60°F), and Annex Q covers low-pressure storage tanks for liquefied hydrocarbon gases at temperatures not lower than –168°C (–270°F).

1.8.3 ASME Section VIII Div (1): *Rules for Construction of Pressure Vessels*

ASME Section VIII Div (1) covers all types of pressure vessels (containers for containment of pressure), either internal or external, applied by an external source, by application of heat from a direct or indirect source, or by any combination thereof. The code lists mandatory requirements, specific prohibitions, and nonmandatory guidance for pressure vessel materials, design, fabrication, examination, inspection, testing, certification, and pressure relief devices. In this case as well, the code does not address all aspects of these activities, and hence it shall not be construed that those aspects that are not specifically addressed are prohibited. In such instances, engineering judgment applied shall be consistent with the underlying philosophy of the code, and such judgments shall never be used to overrule mandatory requirements or specific prohibitions.

The scope of ASME Section VIII Div (1) defines the coverage in detail, by excluding vessels of certain types and pressures. However, on a broader perspective, vessels for containing internal or external pressures above 15 psi (100 kPa) and with dimensions (inside diameter, width, height, or cross-section diagonal) exceeding 6 in. (152 mm) are generally covered by this code.

In addition, ASME Section VIII Div (2) and Div (3) are also available, dealing with pressure vessels, obviously for higher pressure applications.

1.9 Layout of API 650 Based on April 2013 Edition

The API 650 standard is organized in ten chapters from scope through material properties (for generally considered materials), design, fabrication, erection, inspection, and testing, including two chapters related to welding procedure qualification/welders qualification and marking and certification of storage tanks. The details covered by API 650 are summarized in the following table for quick understanding.

Chapter Number	Topic	Details of Topics Covered
1	Scope	General requirements, limitations, responsibilities, documentation requirements, and formulas
2	Normative references	
3	Terms and definitions	
4	Materials	General requirements for plates, sheets, structural shapes, piping and forgings, flanges bolting, welding electrodes, and gaskets
5	Design	Joints, design considerations, special considerations, bottom plates, annular bottom plates, shell design, shell openings, shell attachments and tank appurtenances, top and intermediate stiffening rings, roofs, wind load on tanks (overturning stability), and tank anchorage
6	Fabrication	General requirements and shop inspection
7	Erection	General requirements, details of welding, examination, inspection, testing, repairs, and repairs to welds and dimensional tolerances
8	Methods of examining joints	Radiographic method, magnetic particle examination, ultrasonic examination, liquid penetrant examination, visual examination, and vacuum testing
9	Welding procedure and welder qualification	Definitions, qualification of welding procedures, qualification of welders, and identification of welded joints
10	Marking	Nameplate, division of responsibility and certification
Annex A **(N)**		Optional design basis for small tanks
Annex AL **(N)**		Aluminum storage tanks

(Continued)

Chapter Number	Topic	Details of Topics Covered
Annex B **(I)**		Recommendations for design and construction of foundations for above ground oil storage tanks
Annex C **(N)**		External floating roofs
Annex D **(I)**		Inquiries and suggestions for change
Annex E **(N)**		Seismic design of storage tanks
Annex EC **(I)**		Commentary on Annex E
Annex F **(N)**		Design of tanks for small internal pressures
Annex G **(N)**		Structurally supported aluminum dome roofs
Annex H **(N)**		Internal floating roofs
Annex I **(N)**		Under-tank leak detection and subgrade protection
Annex JA **(N)**		Shop-assembled storage tanks
Annex K **(I)**		Sample application of the variable-design-point method to determine shell-plate thickness
Annex L **(N)**		API STD 650 storage tank data sheet
Annex M **(N)**		Requirements for tanks operating at elevated temperatures
Annex N **(N)**		Use of new materials that are not identified
Annex O **(I)**		Recommendations for under-bottom connections
Annex P **(N)**		Allowable external loads on tank shell openings
Annex S **(N)**		Austenitic stainless steel storage tanks
Annex SC **(N)**		Stainless and carbon steel mixed materials storage tanks
Annex T **(I)**		NDE requirements summary
Annex U **(N)**		Ultrasonic examination in lieu of radiography
Annex V **(N)**		Design of storage tanks for external pressure
Annex W **(I)**		Commercial and documentation recommendations
Annex X **(N)**		Duplex stainless steel storage tanks
Annex Y **(I)**		API monogram
	Figures	
	Tables	

Note: (N) = normative, (I) = informative.

2

Classification of Storage

2.1 Some Basics about Flammable and Combustible Liquids

Flammability of a liquid, which is its ability to produce *ignitable* vapors, is the fundamental hazardous property of flammable and combustible liquid. The properties that have an influence on flammability are shown in the following table.

Property	Description
Vapor pressure and boiling point	—
Flash point	Temperature at which vapors above a liquid's surface can be ignited
Fire point	—
Auto-ignition temperature	Otherwise called self-ignition temperature
Vapor-air density	—
Liquid density	—
Water miscibility	Water miscible liquids are a firefighting challenge

2.2 Classification of Storage (NFPA 30)

The National Fire Protection Association (NFPA), in its code for flammable and combustible liquids (NFPA 30), classifies liquids within its ambit generally as either a flammable liquid or a combustible liquid, and they are defined and classified as follows.

Broadly, liquids that have a closed-cup flash point at or above 37.8°C (100°F) are classified as *combustible liquids*, whereas liquids that have a closed-cup flash point below 37.8°C (100°F) are classified as *flammable liquids*. These broader classifications are further categorized based on finer characteristics, as shown in the following table. The table also provides a few commercially known liquids as well for easy understanding against each subcategory.

Classification		Flash Point	Boiling Point	Examples of Fluids
FLAMMABLE LIQUIDS	IA	< 22.8°C (73°F)	< 37.8°C (100°F)	Diethyl ether, ethylene oxide, some light crude oils
	IB	< 22.8°C (73°F)	≥ 37.8°C (100°F)	Motor and aviation gasoline, toluene, lacquers, lacquer thinner
	IC	≥ 22.8°C ≤ 37.8°C (≥ 73°F ≤ 100°F)		Xylene, some paints, some solvent-based cements
COMBUSTIBLE LIQUIDS	II	≥ 37.8°C < 60°C (≥ 100°F < 140°F)		Diesel fuel, paint thinner
	III	≥ 60°C (≥ 140°F)		
	IIIA	≥ 60°C < 93.0°C (≥ 140°F < 200°F)		Home heating oils
	IIIB	≥ 93°C (≥ 200°F)		Cooking oils, lubricating oils, motor oil

2.3 Classification of Storage (Institute of Petroleum)

The Institute of Petroleum uses a slightly different methodology to classify crude oil and its derivatives, which are potentially hazardous materials. The degree of the hazard is determined essentially by volatility and flash point.

Class		Fluid
Class 0		Liquefied petroleum gases (LPG)
Class I		Liquids that have flash points below 21°C (69.8°F)
Class II	(1)	Liquids that have flash points from 21°C (69.8°F) up to and including 55°C (131°F) handled, below flash point
Class II	(2)	Liquids that have flash points from 21°C (69.8°F) up to and including 55°C (131°F) handled, at or above flash point
Class III	(1)	Liquids that have flash points above 55°C (131°F) up to and including 100°C (212°F) handled, below flash point
Class III	(2)	Liquids that have flash points above 55°C (131°F) up to and including 100°C (212°F) handled, above flash point
Unclassified		Liquids with flash points above 100°C (212°F)

The purpose of these classifications is to arrive at specific requirements for storage facilities with regard to the health and safety of operating personnel and inhabitants within the vicinity, based on the associated hazards of each of the products stored. Since hazard characteristics vary with physical properties of stored fluid, providing standard safeguards for all types of storages might not be economical. Therefore, NFPA and the Institute of Petroleum have developed criteria for storage tanks based on the nature of the fluid stored and the classifications as shown in the above table. Usually, a group of fluids is categorized under each class considering the fluids' physical and chemical characteristics, to ward off complications in formulating requirements for each and every fluid.

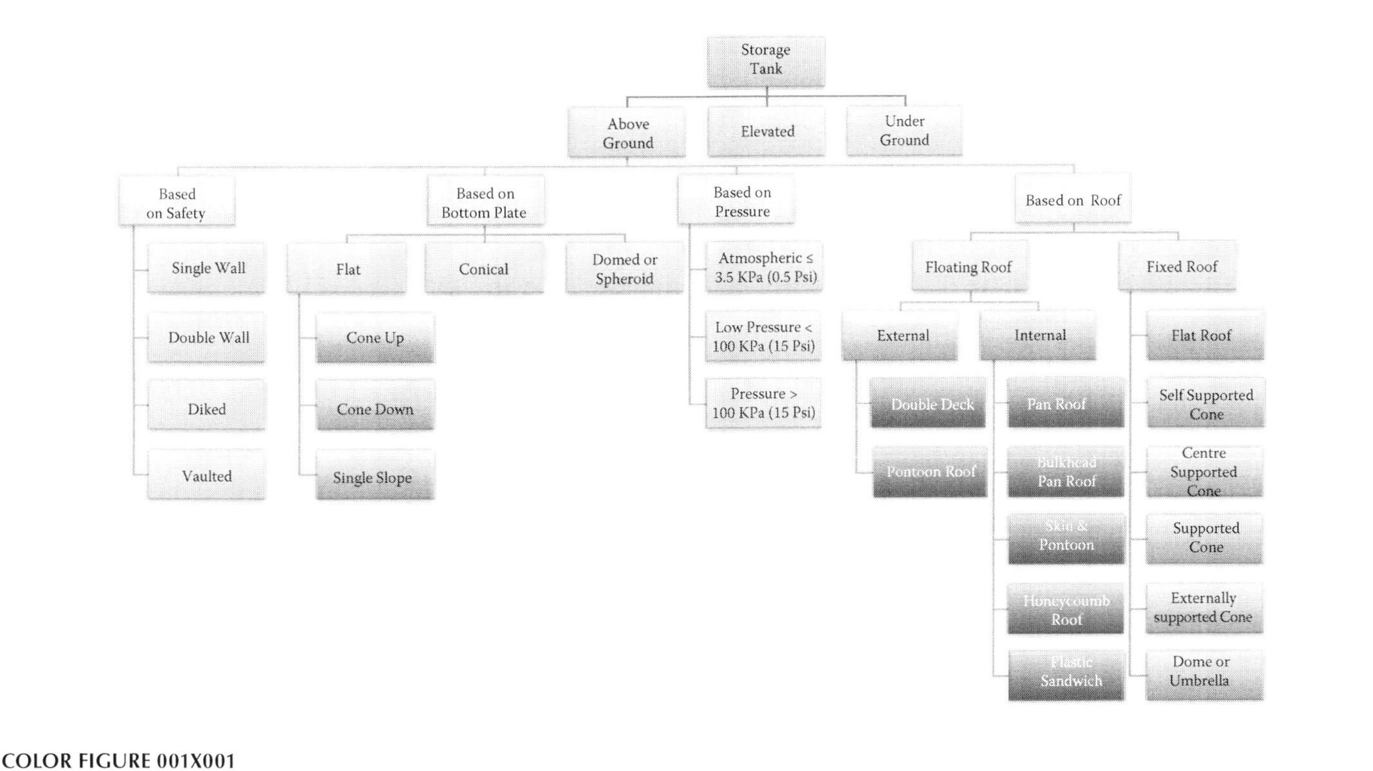

COLOR FIGURE 001X001

COLOR FIGURE 003X001

COLOR FIGURE 003X002

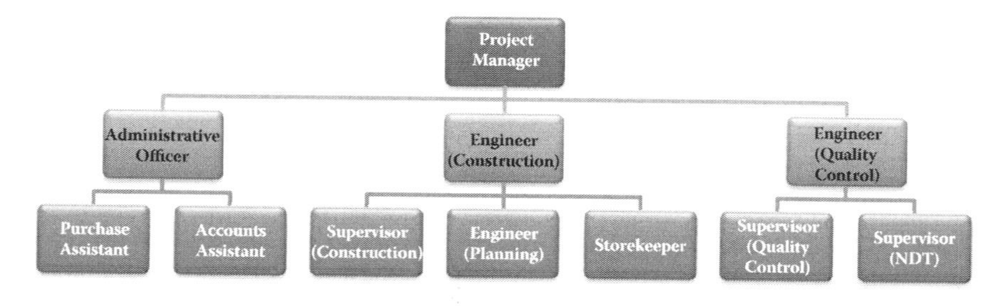

COLOR FIGURE 006X001

COLOR FIGURE 007X001

COLOR FIGURE 007X002

COLOR FIGURE 007X003

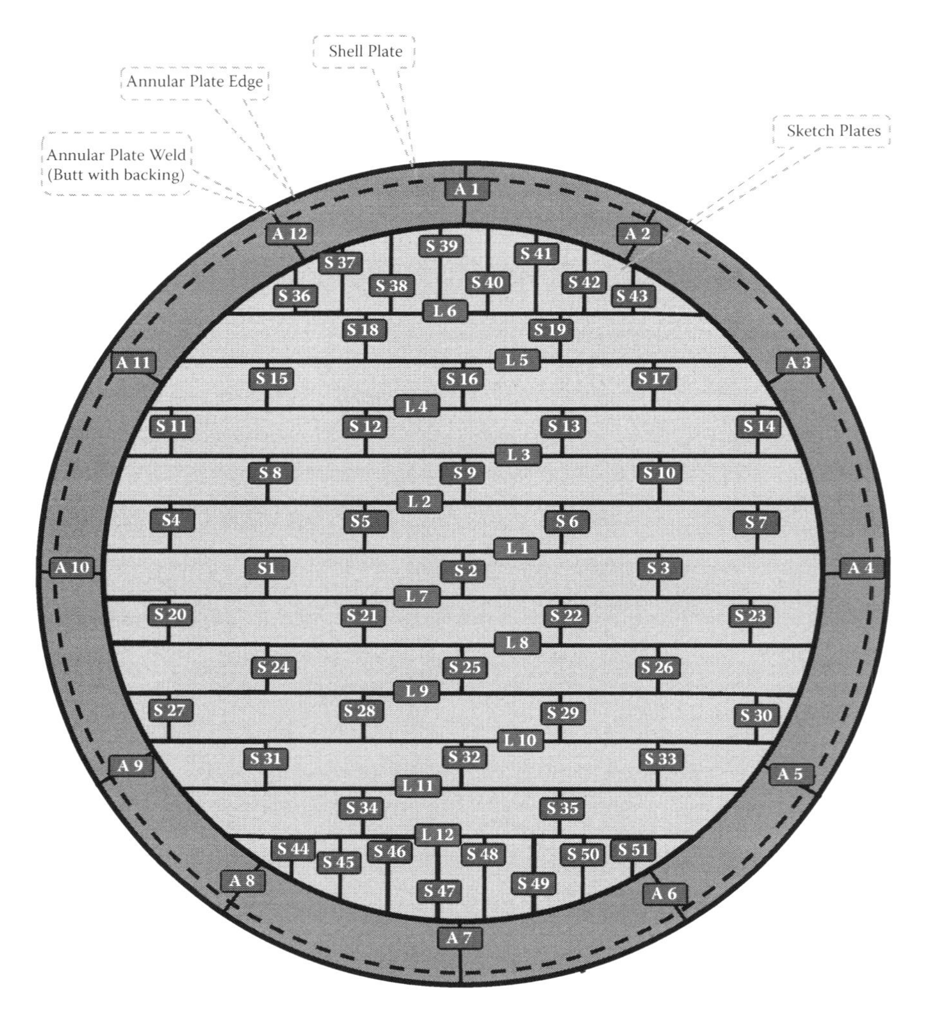

COLOR FIGURE 7.6
Typical layout of annular, sketch, and bottom plates.

COLOR FIGURE 007X009

Shell Cutting Plan with Weld Map

Diameter	Height	Plate size	Type of joining of shells
38,000 mm	20,000 mm	12,000 mm × 2500 mm	Inside flush

Shell thickness	Shell 1	Shell 2	Shell 3	Shell 4	Shell 5	Shell 6	Shell 7	Shell 8
	22 mm	20 mm	18 mm	12 mm	10 mm	8 mm	8 mm	8 mm

H8	S8 V1	S8 V2	S8 V3	S8 V4	S8 V5	S8 V6	S8 V7	S8 V8	S8 V9	S8 V10
	S7 V1	S7 V2	S7 V3	S7 V4	S7 V5	S7 V6	S7 V7	S7 V8	S7 V9	S7 V10 H7
H6	S6 V1	S6 V2	S6 V3	S6 V4	S6 V5	S6 V6	S6 V7	S6 V8	S6 V9	S6 V10
	S5 V1	S5 V2	S5 V3	S5 V4	S5 V5	S5 V6	S5 V7	S5 V8	S5 V9	S5 V10 H5
H4	S4 V1	S4 V2	S4 V3	S4 V4	S4 V5	S4 V6	S4 V7	S4 V8	S4 V9	S4 V10
	S3 V1	S3 V2	S3 V3	S3 V4	S3 V5	S3 V6	S3 V7	S3 V8	S3 V9	S3 V10 H3
H2	S2 V1	S2 V2	S2 V3	S2 V4	S2 V5	S2 V6	S2 V7	S2 V8	S2 V9	S2 V10
	S1 V1	S1 V2	S1 V3	S1 V4	S1 V5	S1 V6	S1 V7	S1 V8	S1 V9	S1 V10 H1

COLOR FIGURE 8.1
Shell cutting plan with weld map.

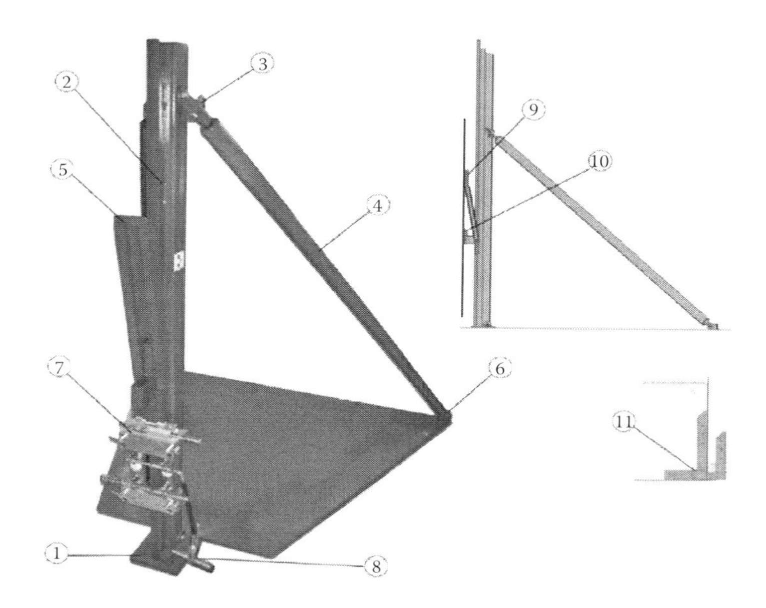

COLOR FIGURE 8.2A
Main components of tank trestle. Drawing Courtesy M/s Bygging Uddemann AB.

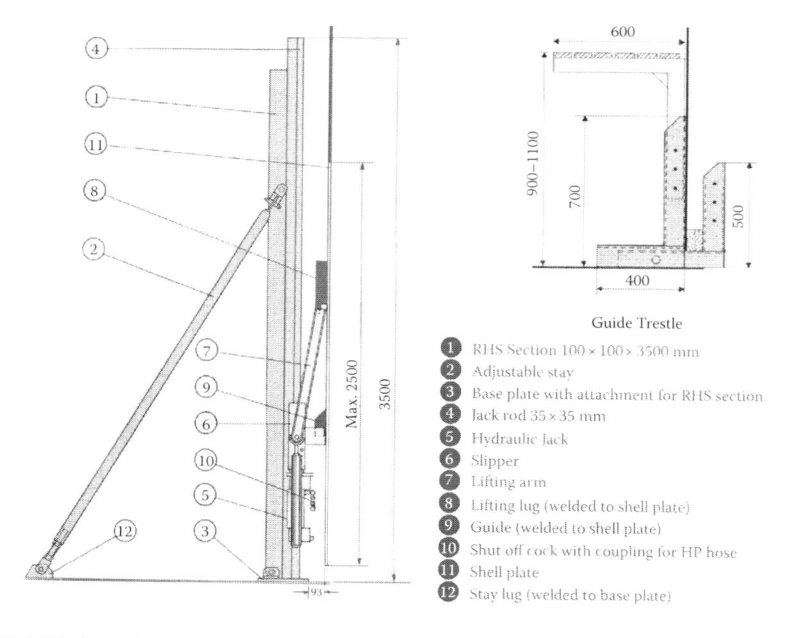

COLOR FIGURE 8.2B
Main components of tank trestles (continued). Drawing Courtesy M/s Bygging Uddemann AB.

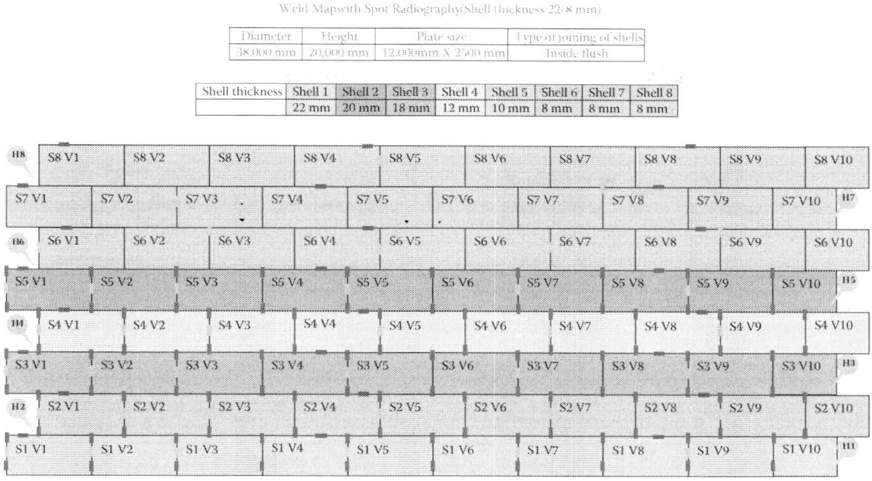

Weld Mapwith Spot Radiography(Shell thickness 22-8 mm)

Diameter	Height	Plate size	Type of joining of shells
38,000 mm	20,000 mm	12,000mm X 2500 mm	Inside flush

Shell thickness	Shell 1	Shell 2	Shell 3	Shell 4	Shell 5	Shell 6	Shell 7	Shell 8
	22 mm	20 mm	18 mm	12 mm	10 mm	8 mm	8 mm	8 mm

H8 S8 V1 S8 V2 S8 V3 S8 V4 S8 V5 S8 V6 S8 V7 S8 V8 S8 V9 S8 V10
S7 V1 S7 V2 S7 V3 S7 V4 S7 V5 S7 V6 S7 V7 S7 V8 S7 V9 S7 V10 H7
H6 S6 V1 S6 V2 S6 V3 S6 V4 S6 V5 S6 V6 S6 V7 S6 V8 S6 V9 S6 V10
S5 V1 S5 V2 S5 V3 S5 V4 S5 V5 S5 V6 S5 V7 S5 V8 S5 V9 S5 V10 H5
H4 S4 V1 S4 V2 S4 V3 S4 V4 S4 V5 S4 V6 S4 V7 S4 V8 S4 V9 S4 V10
S3 V1 S3 V2 S3 V3 S3 V4 S3 V5 S3 V6 S3 V7 S3 V8 S3 V9 S3 V10 H3
H2 S2 V1 S2 V2 S2 V3 S2 V4 S2 V5 S2 V6 S2 V7 S2 V8 S2 V9 S2 V10
S1 V1 S1 V2 S1 V3 S1 V4 S1 V5 S1 V6 S1 V7 S1 V8 S1 V9 S1 V10 H1

COLOR FIGURE 11.1A
Refer to Table 11.1.

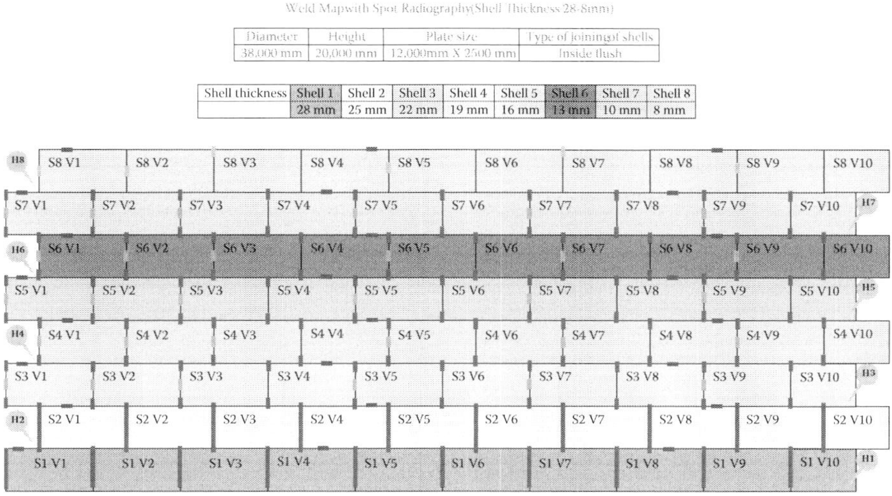

Weld Mapwith Spot Radiography(Shell Thickness 28-8mm)

Diameter	Height	Plate size	Type of joiningof shells
38,000 mm	20,000 mm	12,000mm X 2500 mm	Inside flush

Shell thickness	Shell 1	Shell 2	Shell 3	Shell 4	Shell 5	Shell 6	Shell 7	Shell 8
	28 mm	25 mm	22 mm	19 mm	16 mm	13 mm	10 mm	8 mm

H8 S8 V1 S8 V2 S8 V3 S8 V4 S8 V5 S8 V6 S8 V7 S8 V8 S8 V9 S8 V10
S7 V1 S7 V2 S7 V3 S7 V4 S7 V5 S7 V6 S7 V7 S7 V8 S7 V9 S7 V10 H7
H6 S6 V1 S6 V2 S6 V3 S6 V4 S6 V5 S6 V6 S6 V7 S6 V8 S6 V9 S6 V10
S5 V1 S5 V2 S5 V3 S5 V4 S5 V5 S5 V6 S5 V7 S5 V8 S5 V9 S5 V10 H5
H4 S4 V1 S4 V2 S4 V3 S4 V4 S4 V5 S4 V6 S4 V7 S4 V8 S4 V9 S4 V10
S3 V1 S3 V2 S3 V3 S3 V4 S3 V5 S3 V6 S3 V7 S3 V8 S3 V9 S3 V10 H3
H2 S2 V1 S2 V2 S2 V3 S2 V4 S2 V5 S2 V6 S2 V7 S2 V8 S2 V9 S2 V10
S1 V1 S1 V2 S1 V3 S1 V4 S1 V5 S1 V6 S1 V7 S1 V8 S1 V9 S1 V10 H1

COLOR FIGURE 11.1B
Refer to Table 11.2.

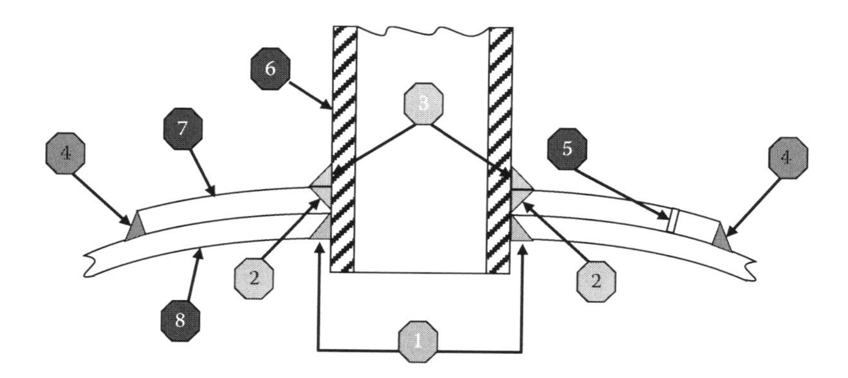

COLOR FIGURE 12.1B
Welds to be inspected.

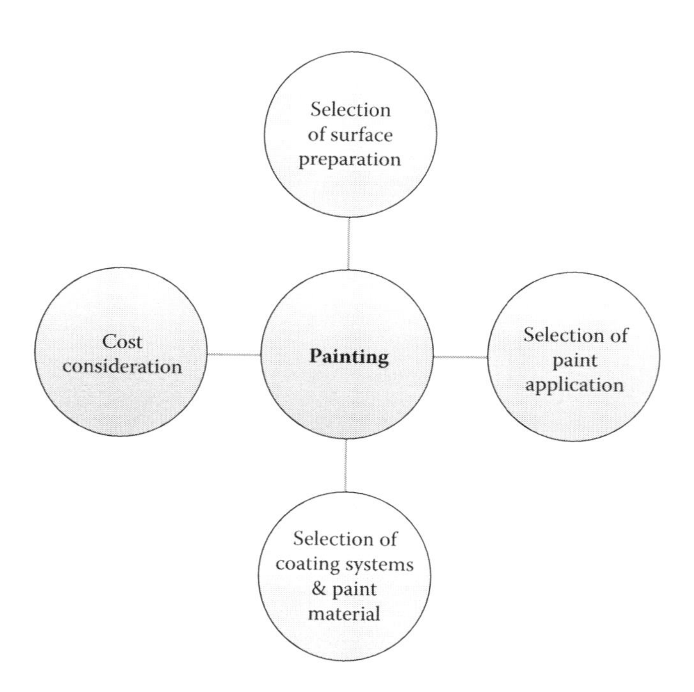

COLOR FIGURE 013X001

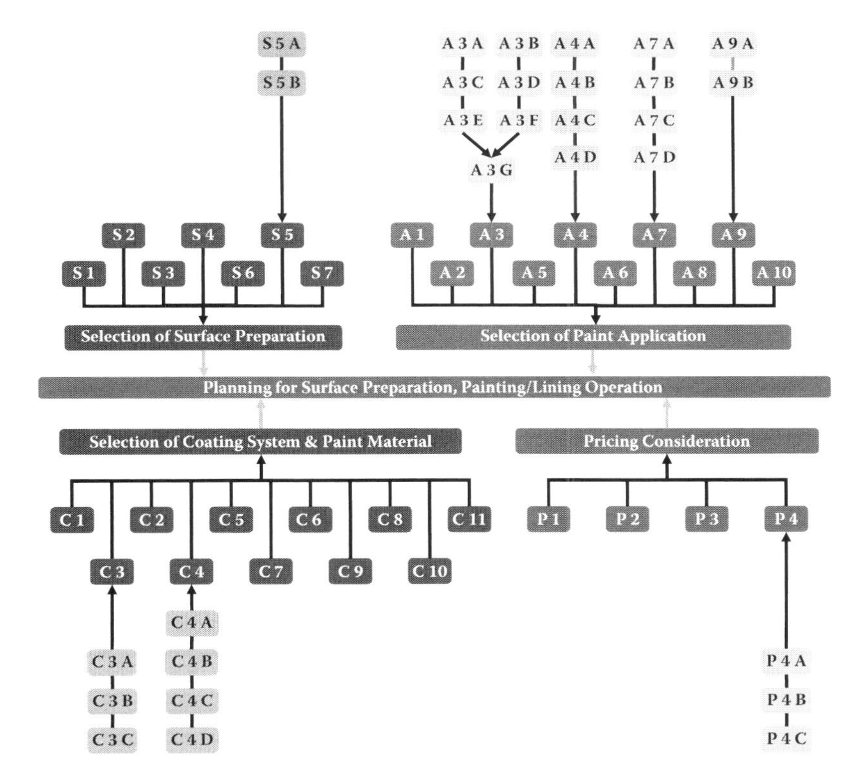

COLOR FIGURE 013X002

Sl. No.	Rust Grade	Description	Picture
1	Rust Grade A	Steel covered completely with adherent mill scale and with if any little rust.	
2	Rust Grade B	Steel surface which has begun to rust and from which the mill scale has begun to flake.	
3	Rust Grade C	Steel surface on which the mill scale has rusted away or from which it can be scrapped, but with little pitting visible to naked eye.	
4	Rust Grade D	Steel surface on which the mill scale has rusted away and on which considerable pitting is visible to naked eye.	

COLOR FIGURE 013X005A

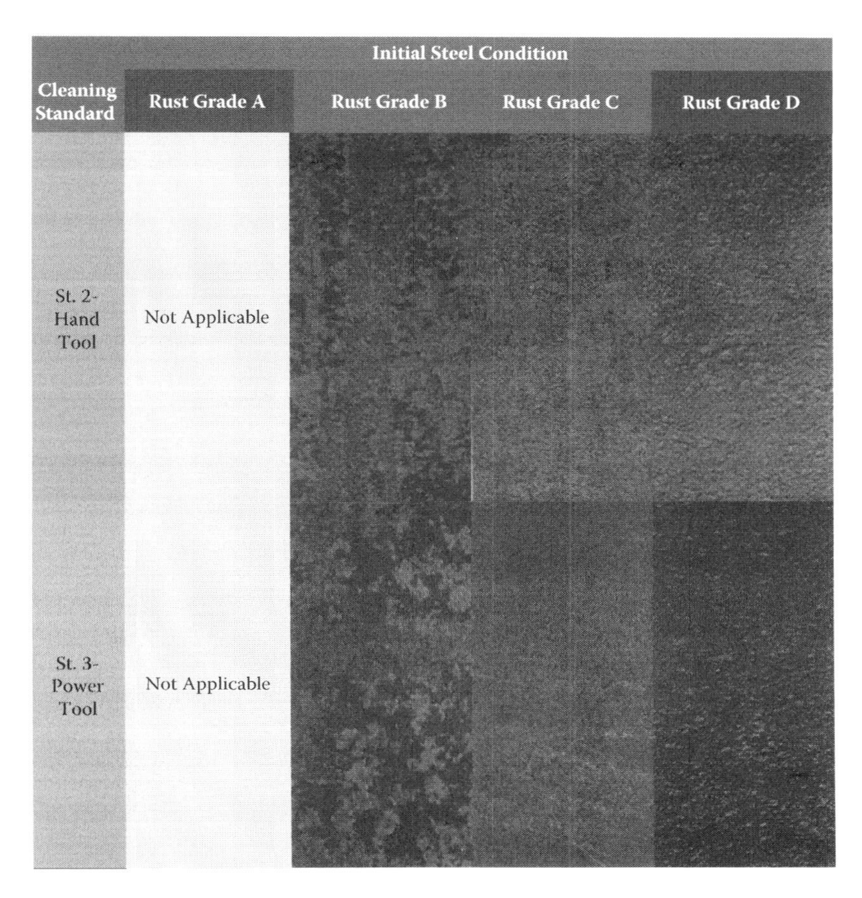

COLOR FIGURE 013X005B

COLOR FIGURE 013X005C

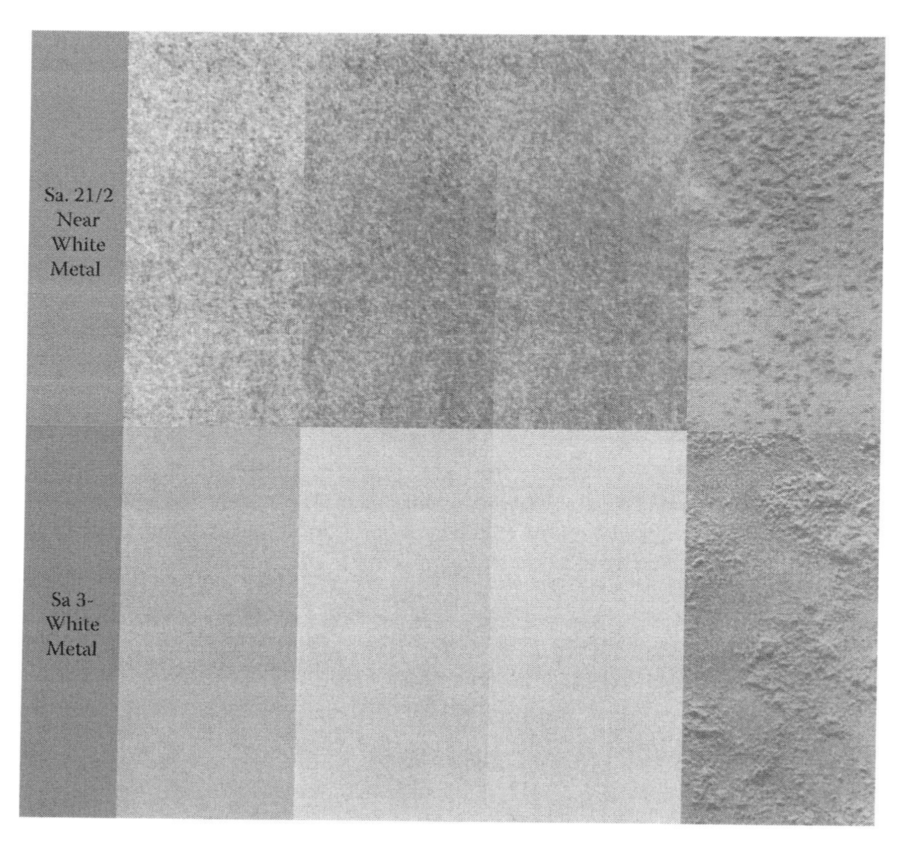

COLOR FIGURE 013X005D

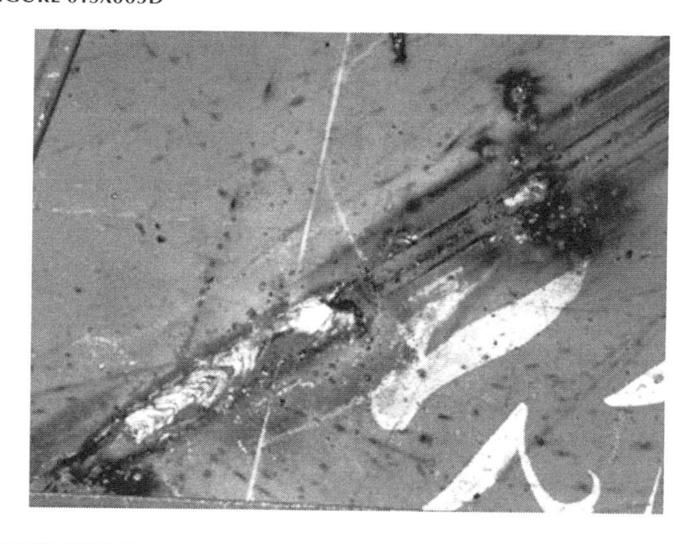

COLOR FIGURE 013X006

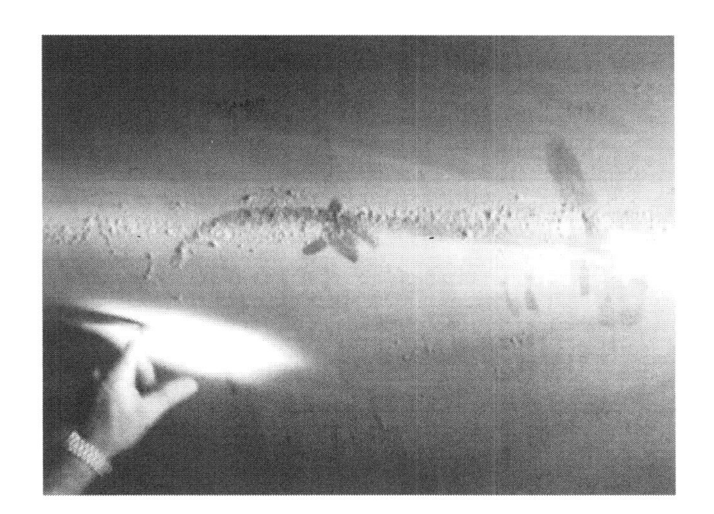

COLOR FIGURE 013X007

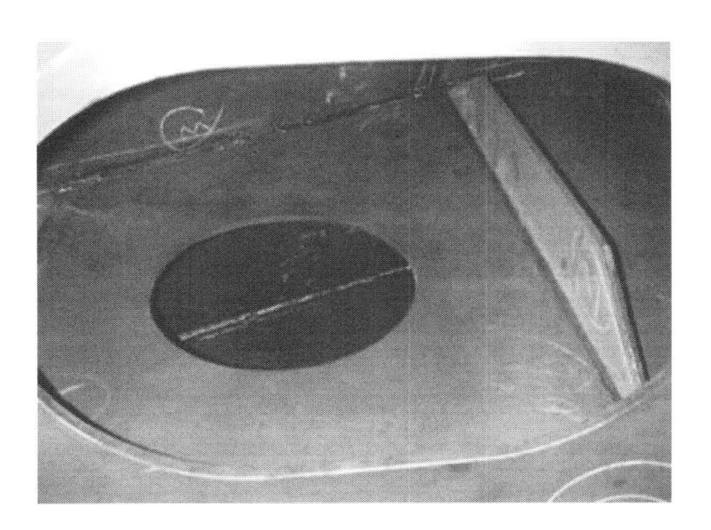

COLOR FIGURE 013X008

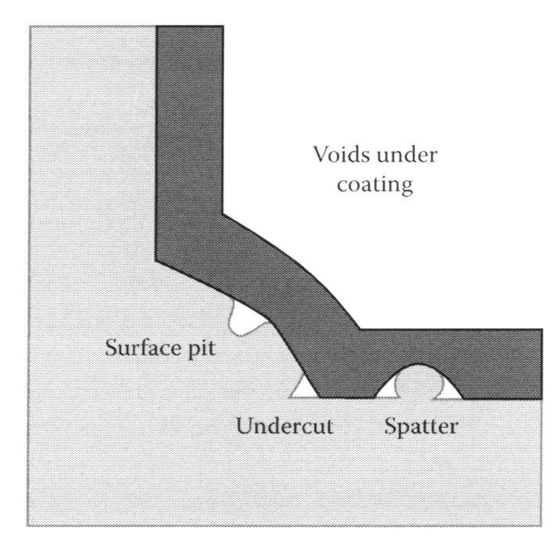

COLOR FIGURE 013X009

3

Tank Farm Layout

3.1 Considerations

A tank farm shall preferably be placed on one or not more than two sides of the process plant area. This arrangement allows adequate safety precautions to be taken and gives the possibility to expand either the tank farm area or the process plant area at any time in the future. Access shall be available on all four sides of each tank bund area, and all roads shall be linked in such a way that access is always possible even when any of the roads is blocked by fire.

For certain liquids, either burial or sun shielding is necessary or desirable. Cooling facilities also may be required in most of the cases. In all cases the layout shall have to satisfy the requirements of local statutory authorities for fire regulations and safety and access requirements. Tanks containing flammable liquids shall be surrounded by bund or dike walls, except those containing fluids with a high flash point at storage temperature (e.g., Class C, asphalt, heavy fuel oils, etc.). In this case, a low wall, 450 mm high, may be desirable in order to control spillage and prevention of pollution.

Areas around tanks can be varied in both size and shape to suit the land available. The type of bund or dike wall also can be varied. In cases where space is available, earth types with side slopes of 1.5 to 2.0 horizontal to 1.0 vertical are cheapest, but they require more space. Where space is a premium, concrete or masonry construction is advantageous. A desirable maximum height for safe access is 2.4 m, but this may be exceeded for very large tanks. Steps shall be provided over the bund with additional emergency exits as required. The minimum effective capacity within the retaining wall shall be equal to 110% that of the largest tank. This rate is based on the assumption that only one tank will fail at a time.

Tanks shall be grouped and bunded so that contents of tanks in one bund shall require only one type of firefighting system. This applies particularly when both water miscible liquids and water immiscible liquids are stored in the same installation. Space shall be allowed for foam or drenching systems. For tanks grouped together, consideration shall be given to a common walkway with not less than two means of escape, depending on the number of tanks served.

When exothermic chemical reactions are possible between stored liquids, tanks shall be segregated from other tanks, and consideration shall be given to increasing the spacing as much as possible.

The layout of storage tanks and related facilities has an impact on general pump arrangements. Pumps related to storages are generally placed in groups to serve one or more tanks and streams. Groups of pumps shall be placed in such a way for easy centralized operation but may require long suction runs of piping and thus prove costly for installation, as well as during operation. Lines carrying hot or flammable material shall be as short as possible.

Pumps shall not be located inside the bund wall around flammable liquid storage. The vertical distance between the tank outlet and the pump suction elevations shall be at least twice the anticipated tank settlement after loading. Steps shall be provided where access routes cross pipework.

Where tanks are provided with heating coils, enough space shall be provided for withdrawal of heating coils for maintenance.

Storage and tank areas need not be provided with lighting if they are not intended to be visited at night by operating staff. If such facilities warrant visit at night because of emergencies, portable lamps with adequate light can be used.

For smaller plants, storage tanks can be located in such a way as to suit the flow of process and hence can be individually located. However, general principles as already narrated shall be adhered to regarding fire, safety, and spillage requirements; safe distances; and so on.

For multiple product storage where products vary according to seasonal or other changes, the layout is important to prevent the accidental mixing of two products and to permit flushing and cleaning of tanks and pipes. In such cases, individual tanks shall not be hard piped to the production plant or to the tanker or container-filling stations. A number of lines shall be provided from the production area to the storage area and from the storage area to the filling point and then connected with a flexible hose according to production requirements.

The layout of tanks, distinct from their spacing, shall always take into consideration the accessibility needed for firefighting and the potential value of a storage tank farm in providing a buffer area between the process plant and public roads, houses, and so on, for environmental reasons. Furthermore, the location of the tank farm relative to process units must be such as to ensure maximum safety from possible incidents. The primary considerations for the layout of storage tank farms can be summarized as follows.

Sl. No.	Description
1	Intertank spacing and separation distances between the tank and the boundary line and the tank and other facilities are of fundamental importance.
2	Suitable roadways shall be available for approach to tank sites by mobile firefighting equipment and personnel.

Sl. No.	Description
3	The fire water system shall be laid out to provide adequate fire protection to all parts of the storage area and transfer facilities.
4	Bund wall and draining of the area surrounding the tanks shall be such that a spillage from any tank can be controlled to minimize subsequent damage to the tank and its contents. They shall also minimize the possibility of other tanks being involved.
5	Tank farms shall preferably not be located at higher levels than process units in the same catchment area.
6	Storage tanks holding flammable liquids shall be installed in such a way that any spill shall not flow toward a process area or any other source of ignition.

3.2 Typical Tank Farms

Completed tank farm.

Tank farm extension work in progress.

3.3 Spacing of Tanks in Tank Farms

Based on the general considerations in deciding a tank farm layout described above, various agencies have put forward recommendations for spacing of tanks (inter tank and between tank and other facilities) within tank farms for process industries. Two such prominent guidelines, widely adopted in the oil and gas industry are NFPA 30 and those of Institute of Petroleum, the gist of which are described/reproduced in the following sections.

3.4 Spacing of Tanks per NFPA 30

NFPA 30 specifies minimum spacing requirements (both between tanks and from other adjacent facilities such as roads, buildings etc.), distances in an elaborate manner based on the type of medium stored, its containment pressure, tank type and it sizes. They are shown as Tables 3.1 to 3.7 below, which are reproductions of Tables 4.3.2.1.1 (a), 4.3.2.1.1 (b), 4.3.2.1.2, 4.3.2.1.3, 4.3.2.1.4, 4.3.2.1.5 and 4.3.2.2.1 of NFPA 30 respectively.

TABLE 3.1

Stable Liquids: Operating Pressure Not to Exceed a Gauge Pressure of 17 kPa (2.5 psi)

		Minimum Distance m (ft)	
Type of Tank	Protection	From Property Line That Is or Can Be Built Upon, Including the Opposite Side of a Public Way[a]	From Nearest Side of Any Public Way or from Nearest Important Building on the Same Property[a]
Floating roof	Protection for exposures[b]	½ × diameter of tank	1/6 × diameter of tank
	None	Diameter of tank but need not to exceed 53.3 m (175 ft)	1/6 × diameter of tank
Vertical with weak roof-to-shell seam	Approved foam or inerting system[c] on tanks not exceeding 50 m (150 ft) in diameter[d]	½ × diameter of tank	1/6 × diameter of tank
	Protection for exposures[b]	Diameter of tank	1/3 × diameter of tank
	None	2 × diameter of tank but need not exceed 106.7 m (350 ft)	
Horizontal and vertical tanks with emergency relief venting to limit pressures to 17 kPa (2.5 psi) (gauge pressure)	Approved inerting system[b] on the tank or approved foam system on vertical tanks	½ × value in Table 3.2	½ × value in Table 3.2
	Protection for exposures[b]	Value in Table 3.2	Value in Table 3.2
	None	2 × value in Table 3.2	Value in Table 3.2
Protected above ground tank	None	½ × value in Table 3.2	½ × value in Table 3.2

Note: For SI units, 1 ft = 0.3 m.

[a] The minimum distance cannot be less than 1.5 m (5 ft).

[b] See definition 3.3.35, Protection for Exposures.

[c] See NFPA 69, Standard on Explosion Prevention Systems.

[d] For tanks over 45 m (150 ft) in diameter, use "Protection for Exposures" or "None," as applicable.

Source: Reproduced from NFPA30 Table 4.3.2.1a.

TABLE 3.2

Reference Table for Use with Tables 3.1, 3.3, and 3.5

	Minimum Distance in m (ft)	
Tank Capacity Range {m^3 (gal)}	**From Property Line That Is or Can Be Built Upon, Including the Opposite Side of a Public Way**	**From Nearest Side of Any Public Way or from Nearest Important Building on the Same Property**
≤ 1.045 (≤ 275)	1.5 (05)	1.5 (5)
> 1.045 ≤ 2.850 (> 275 ≤ 750)	3.0 (10)	1.5 (5)
> 2.850 ≤ 45.60 (> 750 ≤ 12,000)	4.5 (15)	1.5 (5)
> 45.60 ≤ 114.0 (> 12,000 ≤ 30,000)	6.0 (20)	1.5 (5)
> 114.0 ≤ 190.0 (> 30,000 ≤ 50,000)	9.0 (30)	3.0 (10)
> 190.0 ≤ 380.0 (> 50,000 ≤ 100,000)	15.0 (50)	4.5 (15)
> 380.0 ≤ 1900.0 (> 100,000 ≤ 500,000)	24.0 (80)	7.5 (25)
> 1900.0 ≤ 3800.0 (> 500,000 ≤ 1,000,000)	30.0 (100)	10.5 (35)
> 3800.0 ≤ 7600.0 (> 1,000,000 ≤ 2,000,000)	40.5 (135)	13.5 (45)
> 7600.0 ≤ 11400.0 (> 2,000,000 ≤ 3,000,000)	49.5 (165)	16.5 (55)
> 11400.0 (> 3,000,000)	52.5 (175)	18.0 (60)

Note: For SI units, 1 ft = 0.3 m; 1 gal = 3.8 l.
Source: NFPA30 Table 4.3.2.1.1b

TABLE 3.3

Stable Liquids Operating Pressure Greater than Gauge Pressure of 17 kPa (2.5 psi)

		Minimum Distance m (ft)	
Type of Tank	**Protection**	**From Property Line That Is or Can Be Built Upon, Including the Opposite Side of a Public Way**	**From Nearest Side of Any Public Way or from Nearest Important Building on the Same Property**
Any type	Protection for exposures[a]	1½ × Table 3.2 value but shall not be less than 7.5 m (25 ft)	1½ × Table 3.2 value but shall not be less than 7.5 m (25 ft)
	None	3 × Table 3.2 value but shall not be less than 7.5 m (25 ft)	1½ × Table 3.2 value but shall not be less than 7.5 m (25 ft)

Note: For SI units, 1 ft = 0.3 m.
[a] See definition 3.3.42, Protection for Exposures.
Source: NFPA30 Table 4.3.2.1.2

TABLE 3.4

Boil Over Liquids

		Minimum Distance m (ft)	
Type of Tank	**Protection**	**From Property Line That Is or Can Be Built Upon, Including the Opposite Side of a Public Way**[a]	**From Nearest Side of Any Public Way or from Nearest Important Building on the Same Property**[a]
Floating roof	Protection for exposures[b]	½ × diameter of tank	1/6 × diameter of tank
	None	Diameter of tank	1/6 × diameter of tank
Fixed roof	Approved foam or inerting system[c]	Diameter of tank	1/6 × diameter of tank
	Protection for exposures[b]	2 × diameter of tank	2/3 × diameter of tank
	None	4 × diameter of tank but need not exceed 105 m (350 ft)	2/3 × diameter of tank

Note: For SI units, 1 ft = 0.3 m.

[a] The minimum distance cannot be less than 1.5 m (5 ft).

[b] See definition 3.3.35, Protection for Exposures.

[c] See NFPA 69, Standard on Explosion Prevention Systems.

Source: NFPA30 Table 4.3.2.1.3.

TABLE 3.5

Unstable Liquids

Type of Tank	Protection	Minimum Distance m (ft)	
		From Property Line That Is or Can Be Built Upon, Including the Opposite Side of a Public Way	From Nearest Side of Any Public Way or from Nearest Important Building on the Same Property
Horizontal and vertical tanks with emergency relief venting to permit pressure not in excess of a gauge pressure of 17 kPa (2.5 psig)	Tank protected with any one of the following: approved water spray, approved inerting,[a] approved insulation and refrigeration, approved barricade	Table 3.2 value but not less than 7.5 m (25 ft)	Not less than 7.5 m (25 ft)
	Protection for exposures[b]	2½ × Table 3.2 value but not less than 15 m (50 ft)	Not less than 15 m (50 ft)
	None	5 × Table 3.2 value but not less than 30 m (100 ft)	Not less than 30 m (100 ft)
Horizontal and vertical tanks with emergency relief venting to permit pressure over a gauge pressure of 17 kPa (2.5 psig)	Tank protected with any one of the following: approved water spray, approved inerting,[a] approved insulation and refrigeration, approved barricade	2 × Table 3.2 value but not less than 15 m (50 ft)	Not less than 15 m (50 ft)
	Protection for exposures[b]	4 × Table 3.2 value but not less than 30 m (100 ft)	Not less than 30 m (100 ft)
	None	8 × Table 3.2 value but not less than 45 m (150 ft)	Not less than 45 m (150 ft)

Note: For SI units, 1 ft = 0.3 m.

[a] See NFPA 69, Standard on Explosion Prevention Systems.

[b] See definition 3.3.35, Protection for Exposures.

Source: NFPA Table 4.3.2.1.4.

TABLE 3.6

Class III B Fluids

Tank Capacity Range {m³ (gal)}	Minimum Distance m (ft)	
	From Property Line That Is or Can Be Built Upon, Including the Opposite Side of a Public Way	From Nearest Side of Any Public Way or from Nearest Important Building on the Same Property
≤ 45.60 (≤ 12,000)	1.5 (5)	1.5 (5)
> 45.60 ≤ 114.00 (> 12,000 ≤ 30,000)	3 (10)	1.5 (5)
> 114.00 ≤ 190.00 (> 30,000 ≤ 50,000)	3 (10)	3 (10)
> 190.00 ≤ 380 00 (> 50,000 ≤ 100,000)	4.5 (15)	3 (10)
> 380.00 (> 100,000)	4.5 (15)	4.5 (15)

Note: For SI units, 1 ft = 0.3 m; 1 gal = 3.8 l.
Source: NFPA 30 Table 4.3.2.1.5.

Tanks storing Class I, Class II, or Class III stable liquids shall be separated by distances given in Table 3.7.

Exception No. 1: Tanks storing crude petroleum that have individual capacities not exceeding 480 m3 (126,000 gal or 3,000 bbl) and that are located at production facilities in isolated locations do not need to be separated by more than 1 m (3 ft).

Exception No. 2: Tanks used only for storing Class III B liquids need not be separated by more than 1 m (3 ft) provided they are not within the same diked area as, or within the drainage path of, a tank storing a Class I or Class II liquid.

TABLE 3.7

Minimum Tank Spacing (Shell to Shell)

Tank Diameter	Floating Roof Tanks	Fixed or Horizontal Tanks	
		Class I or II Liquids	Class III A Liquids
All tanks not over 45 m (150 ft) in diameter	1/6 × sum of adjacent tank diameters but not less than 1 m (3 ft)	1/6 × sum of adjacent tank diameters but not less than 1 m (3 ft)	1/6 × sum of adjacent tank diameters but not less than 1 m (3 ft)
Tanks larger than 45 m (150 ft) in diameter			
If remote impounding is provided in accordance with NFPA 30 Table 4.3.2.3.1	1/6 × sum of adjacent tank diameters	1/4 × sum of adjacent tank diameters	1/6 × sum of adjacent tank diameters
If diking is provided in accordance with 4.3.2.3.2	1/4 × sum of adjacent tank diameters	1/3 × sum of adjacent tank diameters	1/4 × sum of adjacent tank diameters

Note: For SI units, 1 ft = 0.3 m.
Source: NFPA 30 Table 4.3.2.2.1.

3.5 Spacing of a Tank for Petroleum Stocks as per the Institute of Petroleum Guidelines

As mentioned earlier, the layout and general design of a petroleum storage installation is based on considerations of safety, operational ease, and environmental conservation. Installations for Class II(1) or Class III(1) petroleum pose a lesser risk than those handling Class I, Class II(2), or Class III(2) fluids.

3.5.1 Storage Classes I, II(2), III(2)

The distances given in Table 3.8 (Reproduced from IP guideline Table 3.1) are normal minimum recommendations applicable to storage of Classes I, II(2), and III(2). Numerical values provided refer to horizontal distances in the plan between the nearest points of the specified features (e.g., storage tanks, filling points, openings in buildings, and boundaries). Special consideration should be given to sites on sloping ground or where high wind speeds prevail. The distances described in the table should be used in conjunction with the appropriate level of fire protection envisaged for the facility.

A group of small tanks referred to in the table is defined as tanks having a diameter of 10 m (30 ft) or less and a height of 14 m (46 ft) or less and may be regarded as one tank when considering tank spacing or bunding. Such small tanks may be placed together in groups, no groups having an aggregate capacity of more than 8,000 m^3 (282,000 ft^3). The distance between individual tanks in the group need to be governed only by constructional and operating convenience.

3.5.2 Storage Classes II(1) and III(1)

Petroleum products stored at installations and depots may be regarded as Class II(1) in temperate climates since they will be below their flash points at ambient temperatures. The spacing of tanks need be governed by constructional and operational convenience only.

However, it is recommended that tanks for Class II products should be spaced from tanks storing Class I products at the distances for fixed roof tanks shown in Table 3.8.

At refineries, products or product component stocks may at times be held at temperatures higher than their flash points. In this case, Class II materials will be classified as Class II(2), and spacing should be in accordance with that shown in Table 3.8.

TABLE 3.8

Location and Spacing for Above Ground Tanks for Product Storage, Classes I, II(2), and III(2)

Tank Type		Factor	Recommended Minimum Distances
Fixed roof, above ground, including those with internal floating roofs, horizontal cylindrical tanks	1	Between tanks within a group of small tanks	Determined solely by construction and maintenance operational convenience
	2	Between groups of small tanks	15 m (45 ft)
	3	Between a group of small tanks and any tank outside the group	Not less than 10 m (30 ft); need not exceed 15 m (45 ft)
	4	Between tanks not being part of a group of small tanks	Half the diameter of the larger tank or diameter of the smaller tank, whichever is less, but in no case less than 10 m (30 ft); need not exceed 15 m (45 ft)
	5	Between a tank and any filling point, filling shed, or building not containing a fixed source of ignition	15 m (45 ft) but in agreement with licensing authority, and where tanks are small, this may be reduced to not less than 6 m (18 ft)
	6	Between a tank and outer boundary or installation, any designated nonhazardous are, or any fixed source of ignition at ground level	15 m (45 ft)
Floating roof	1	Within a group of small tanks	As above for fixed roof tanks[a,b]
	2	Between two floating roof tanks[c,d]	10 m (30 ft) for tanks up to and including 45 m (135 ft) 15 m (45 ft) for tanks above 45 m (135 ft). The size of the larger tank should govern the spacing. For crude oil, not less than 10 m (30 ft), but a spacing of 0.3 D should be considered, with no upper limit.
	3	Between a floating roof tank and a fixed roof tank	Half the diameter of the larger tank or diameter of the smaller tank, whichever is less, but in no case less than 10 m (30 ft); need not exceed 15 m (45 ft)
	4	Between a floating roof tank and any filling point, filling shed, or building not containing a possible source of ignition	10 m (30 ft)
	5	Between a floating roof tank and outer boundary or installation, any designated nonhazardous are, or any fixed source of ignition at ground level	15 m (45 ft)

Note: a. For tanks greater than 18 m (55 ft) in height, it may be necessary to consider whether the distances listed in this table should be increased to take account of the height of the tank.

b. Fixed roof tanks fitted with internal floating roofs may be considered as fixed roof tanks for the purpose of tank location and spacing.

c. For tanks greater than 18 m (55 ft) in height, see Note 1.

d. Floating roof tanks fitted with external metal domed roofs extending over the entire roof area may be considered as fixed roof tanks for the purpose of tank location and spacing.

Source: Table 3.1 of IP Guidelines.

While Table 3.8 provides guidance on the minimum tank spacing for Classes I, II, and III(2) storage facilities, the following points also shall be considered while applying the same:

1. Tanks of diameter up to 10 m (30 ft) are classed as small tanks.
2. Small tanks may be sited together in groups, no group having an aggregate capacity of more than 8,000 m^3 (282,500 ft^3). Such a group may be regarded as one tank.
3. Where future changes of service of a storage tank are anticipated, the layout and spacing should be designed for the most stringent case.
4. For reasons of firefighting access, there should be no more than two rows of tanks between adjacent access roads.
5. Fixed roof tanks with internal floating covers should be treated for spacing purposes as fixed roof tanks.
6. Where fixed roof and floating roof tanks are adjacent, spacing should be on the basis of the tank(s) with the most stringent conditions.
7. Where tanks are erected on compressible soils, the distance between adjacent tanks should be sufficient to avoid excessive distortion. This can be caused by additional settlements of the ground where the stressed soil zone of one tank overlaps that of the adjacent tank.
8. For Class III(1) and unclassified petroleum stocks, spacing of tanks is governed only by constructional and operational convenience. However, the spacing of Class III(1) tankage from Classes I, II, and III(2) tankage is governed by the requirements for the latter.

3.6 API Tank Sizes for Layout Purpose (Based on API 650)

For layout purposes, based on API 650, tank sizes are arrived at as follows:

Capacity Approximately		Diameter (Meters)	Height (Meters)
Meter^3	U.S. Barrels		
75	500	4.6	4.9
150	1.000	6.4	4.9
225	1.500	6.4	7.3
300	2.000	7.6	7.3
450	3.000	9.2	7.3
600	4.000	9.2	9.3
750	5.000	9.2	12.2
900	6.000	9.2	14.6
1,050	7.000	12.2	9.9
1,350	9.000	12.2	12.2
1,500	10.000	12.8	12.2
1,800	12.000	12.8	14.6
2,250	15.000	14.6	14.6
3,000	20.000	18.3	12.2
4,500	30.000	22.3	12.2
6,000	40.000	26.0	12.2
7,500	50.000	27.5	14.6
12,000	90.000	36.6	12.2
15,000	100.000	41.0	12.2
18,000	120.000	41.0	14.6
21,000	140.000	49.8	12.2
27,000	180.000	54.9	12.2
30,000	200.000	54.9	14.6
45,000	300.000	61.0	17.0
60,000	450.000	73.2	17.0
90,000	600.000	91.5	14.6
100,000	800.000	105.0	14.6

3.7 Summary of Refinery Storage Tanks

		Conservation-Type Storage Tanks		
Characteristics	Standard Storage Tanks	Floating Roof	Variable Vapor Space	Pressure Storage
Evaporation losses	High	Significantly reduced	Significantly reduced	Prevented or eliminated
Operating conditions	Recommended for liquids whose vapor pressure is atmospheric or below at storage conditions (vented)	Allow no vapor space above liquid level (no venting)	Allow air vapor mixture to change volume at constant or variable pressure (no venting)	Allow the pressure in vapor space to build up. Tanks are capable of withstanding the maximum pressure without venting.
Subclassifications	Rectangular Cylindrical Vertical Horizontal			Low pressure normally designed for 17–34 kPa (2.5–5 psig) up to 103 kPa (15 psig), High-pressure storage, 207–1,379 kPa (30–200 psig)
Typical types	Cone roof vertical (cylindrical tanks)	Floating roof, Wiggins–Hideck type	Lifter roof tanks, Wiggins dry seal type	Spheroids and hemispheroids for low-pressure storage; spheres for high-pressure storage
Applications	Heavy refinery products	Sour crude oils, light crude oils, light products	Light refinery products and distillates	Spheroids are used to store aviation, motor, jet fuels; spheres are used to store natural gasoline and LPG

3.8 Use of Floating and Fixed Roof Tanks

3.8.1 Floating Roof Tanks

Floating roof tanks are generally used for volatile liquids to minimize product loss and for safety and environmental reasons. Furthermore, there is a preference for floating roof tanks over fixed roof tanks as the size of the tank increases, as the vapor pressure of the stored liquid increases, and when the flash point is below the storage temperature.

The roof in such cases consists of an arrangement of buoyant compartments (pontoons) that floats on the liquid. The gap between the floating roof and the tank is sealed by mechanical means or by tubular-type seals. The roof is provided with support legs that can be adjusted to hold it in either of two positions, the upper position high enough to permit access for tank cleaning and maintenance. The lower position should keep the roof just above the inlet and outlet nozzles, drainage system, and other accessories located at the tank's bottom.

Floating roof tanks are normally equipped with rolling ladders. The tank shell is earthed, and the roof and all fittings, such as rolling ladders, are adequately bonded to the shell as a protection against lightning. All internals such as gauge floats, cables, mixers, and so on also shall be suitably earthed to prevent accumulation of electric charge.

3.8.2 Fixed Roof Tanks

Fixed roof tanks are generally used in refineries where the product stored does not readily vaporize at ambient or stored temperature conditions. The size of the tank and flash point of the product stored shall also influence the choice of tank. For any type of tank, during operation, the space between the roof and the liquid is filled with the vapor of the liquid stored. Depending on materials to be stored, fixed roof tanks can be designed for storage at atmospheric pressure in which case they are equipped with open vents. API 650 covers storage tanks with a maximum internal pressure above 17 kPa (2.5 psig). Weak shell-to-roof welds can be incorporated to give protection to the tank shell in the event of excessive internal pressure. Fixed roof tanks shall be adequately earthed as a protection against lightning as well.

3.8.3 Considerations to Decide between Floating Roof Type and Fixed Roof Type

The following shows a decision tree to select between floating roof tanks and fixed roof tanks and also the criteria for selecting the type as a rule of thumb based on life-cycle costing.

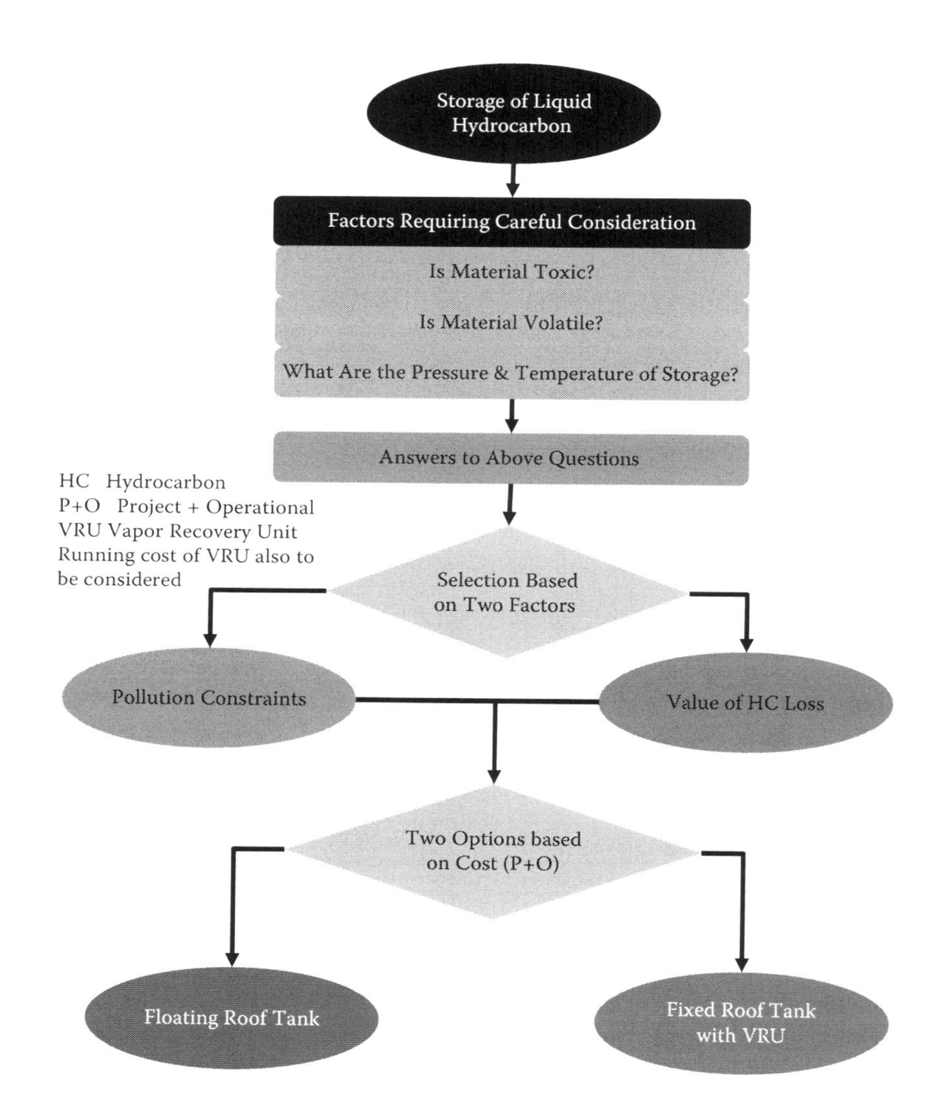
Storage of Liquid Hydrocarbon
Factors Requiring Careful Consideration
Is Material Toxic?
Is Material Volatile?
What Are the Pressure & Temperature of Storage?
Answers to Above Questions
HC Hydrocarbon
P+O Project + Operational
VRU Vapor Recovery Unit
Running cost of VRU also to be considered
Selection Based on Two Factors
Pollution Constraints
Value of HC Loss
Two Options based on Cost (P+O)
Floating Roof Tank
Fixed Roof Tank with VRU

3.8.4 Tank Type Selection Guidelines

	Tank Diameter (m)																							
	3	4	6	8	10	12.5	15	17.5	20	22.5	25	27.5	30	33	36	39	42	45	48	54	60	66	72	80
Type of tank	NPC, LPC, HPC, and OT						NPC, LPC, HPC, NPD, LPD, HPD, OT, and OT(F)			NPC, LPC, NPD, LPD, OT, and OT(F)							NPC, OT, and OT(F)			NPC, OT, and OT(F)		OT and OT(F)		
Type of roofs	*Cone Roof Tanks*																							
	Radial rafters						Internal trusses													IT (special design)		Not available		
	Dome Roof Tanks																							
	Not available						Internal trusses										Not available							
	Floating Roof Tanks																							
	Under special circumstances only						Pontoon type. Double deck type may be used under special circumstances.														Double deck type			
	Fixed Roof Tanks with Floating Deck Inside																							
	May be used under special circumstances																Not recommended							
Shell plates	Width of plate minimum 1.5 m and maximum 2.0 m						Width of plate minimum 2.0 m and maximum 3.0 m																	
	Bottom shell course 8.0 mm minimum Remaining shell courses 6.0 mm minimum													8.0 mm minimum						10.0 mm minimum				
Bottom plates	Minimum thickness 6.0 mm																							
Bottom annular plates	Minimum thickness 8.0 mm						Thickness 10.0 mm, 12.5 mm, or 15 mm																	

Note: NPC = nonpressure cone, LPC = low-pressure cone, HPC = high-pressure cone, NPD = nonpressure dome, LPD = low-pressure dome, HPD = high-pressure dome, OT = open tank, OT(F) = open tank (floating), IT = internal truss.

3.9 Storage Tank Capacity Chart

Storage Tank Capacity Chart																														
Height Meter	Tank Diameter (Meters)																													
	3.0	4.0	6.0	8.0	10.0	12.5	15.0	17.5	20.0	22.5	25.0	27.5	30.0	33.0	36.0	39.0	40.0	45.0	48.0	51.0	54.0	57.0	60.0	66.0	72.0	78.0	84.0	90.0	96.0	102.0
1.0	7	13	28	50	79	123	177	241	314	398	491	594	707	855	1,018	1,195	1,257	1,590	1,810	2,043	2,290	2,552	2,827	3,421	4,072	4,778	5,542	6,362	7,238	8,171
2.0	14	25	57	101	157	245	353	481	628	795	982	1,188	1,414	1,711	2,036	2,389	2,513	3,181	3,619	4,086	4,580	5,104	5,655	6,842	8,143	9,557	11,084	12,723	14,476	16,343
3.0	21	38	85	151	236	368	530	722	942	1,193	1,473	1,782	2,121	2,566	3,054	3,584	3,770	4,771	5,429	6,128	6,871	7,655	8,482	10,264	12,215	14,335	16,625	19,085	21,715	24,514
4.0	28	50	113	201	314	491	707	962	1,257	1,590	1,963	2,376	2,827	3,421	4,072	4,778	5,027	6,362	7,238	8,171	9,161	10,207	11,310	13,685	16,286	19,113	22,167	25,447	28,953	32,685
5.0	35	63	141	251	393	614	884	1,203	1,571	1,988	2,454	2,970	3,534	4,276	5,089	5,973	6,283	7,952	9,048	10,214	11,451	12,759	14,137	17,106	20,358	23,892	27,709	31,809	36,191	40,856
6.0	42	75	170	302	471	736	1,060	1,443	1,885	2,386	2,945	3,564	4,241	5,132	6,107	7,168	7,540	9,543	10,857	12,257	13,741	15,311	16,965	20,527	24,429	28,670	33,251	38,170	43,429	49,028
7.0		88	198	352	550	859	1,237	1,684	2,199	2.783	3,436	4,158	4,948	5,987	7,125	8,362	8,796	11,133	12,667	14,300	16,032	17,862	19,792	23,948	28,501	33,449	38,792	44,532	50,668	57,199
8.0		101	226	402	628	982	1,414	1,924	2,513	3,181	3,927	4,752	5,655	6,842	8,143	9,557	10,053	12,723	14,476	16,343	18,322	20,414	22,619	27,370	32,572	38,227	44,334	50,894	57,906	65,370
9.0			254	452	707	1,104	1,590	2,165	2,827	3,578	4,418	5,346	6,362	7,698	9,161	10,751	11,310	14,314	16,286	18,385	20,612	22,966	25,447	30,791	36,644	43,005	49,876	57,256	65,144	73,542
10.0			283	503	785	1,227	1,767	2,405	3,142	3,976	4,909	5,940	7,069	8,553	10,179	11,946	12,566	15,904	18,096	20,428	22,902	25,518	28,274	34,212	40,715	47,784	55,418	63,617	72,382	81,713
11.0				553	864	1,350	1,944	2,646	3,456	4,374	5,400	6,534	7,775	9,408	11,197	13,140	13,823	17,495	19,905	22,471	25,192	28,069	31,102	37,633	44,787	52,562	60,959	69,979	79,621	89,884
12.0				603	942	1,473	2,121	2,886	3,770	4,771	5,890	7,127	8,482	10,264	12,215	14,335	15,080	19,085	21,715	24,514	27,483	30,621	33,929	41,054	48,858	57,340	66,501	76,341	86,859	98,055
13.0					1,021	1,595	2,297	3,127	4,084	5,169	6,381	7,721	9,189	11,119	13,232	15,530	16,336	20,676	23,524	26,557	29,773	33,173	36,757	44,476	52,930	62,119	72,043	82,702	94,097	106,227
14.0					1,100	1,718	2,474	3,367	4,398	5,567	6,872	8,315	9,896	11,974	14,250	16,724	17,593	22,266	25,334	28,599	32,063	35,725	39,584	47,897	57,001	66,897	77,585	89,064	101,335	114,398
15.0					1,178	1,841	2,651	3,608	4,712	5,964	7,363	8,909	10,603	12,829	15,268	17,919	18,850	23,856	27,143	30,642	34,353	38,276	42,412	51,318	61,073	71,675	83,127	95,426	108,573	122,569
16.0					1,257	1,963	2,827	3,848	5,027	6,362	7,854	9,503	11,310	13,685	16,286	19,113	20,106	25,447	28,953	32,685	36,644	40,828	45,239	54,739	65,144	76,454	88,668	101,788	115,812	130,741
17.0						2,086	3,004	4,089	5,341	6,759	8,345	10,097	12,017	14,540	17,304	20,308	21,363	27,037	30,762	34,728	38,934	43,380	48,066	58,160	69,216	81,232	94,210	108,149	123,050	138,912
18.0						2,209	3,181	4,330	5,655	7,157	8,836	10,691	12,723	15,395	18,322	21,503	22,619	28,628	32,572	36,771	41,224	45,932	50,894	61,581	73,287	86,011	99,752	114,511	130,288	147,083
19.0						2,332	3,358	4,570	5,969	7,555	9,327	11,285	13,430	16,251	19,340	22,697	23,876	30,218	34,382	38,814	43,514	48,483	53,721	65,003	77,359	90,789	105,294	120,873	137,526	155,254
20.0						2,454	3,534	4,811	6,283	7,952	9,817	11,879	14,137	17,106	20,358	23,892	25,133	31,809	36,191	40,856	45,804	51,035	56,549	68,424	81,430	95,567	110,835	127,235	144,765	163,426
21.0							3,711	5,051	6,597	8,350	10,308	12,473	14,844	17,961	21,375	25,086	26,389	33,399	38,001	42,899	48,095	53,587	59,376	71,845	85,502	100,346	116,377	133,596	152,003	171,597
22.0							3,888	5,292	6,912	8,747	10,799	13,067	15,551	18,817	22,393	26,281	27,646	34,989	39,810	44,942	50,385	56,139	62,204	75,266	89,573	105,124	121,919	139,958	159,241	
23.0							4,064	5,532	7,226	9,145	11,290	13,661	16,258	19,672	23,411	27,476	28,903	36,580	41,620	46,985	52,675	58,690	65,031	78,687	93,645	109,902	127,461	146,320	166,479	
24.0							4,241	5,773	7,540	9,543	11,781	14,255	16,965	20,527	24,429	28,670	30,159	38,170	43,429	49,028	54,965	61,242	67,858	82,109	97,716	114,681	133,002			
25.0							4,418	6,013	7,854	9,940	12,272	14,849	17,671	21,382	25,447	29,865	31,416	39,761	45,239	51,071	57,256	63,794	70,686	85,530	101,788	119,459	138,544			

4

Tank Design

4.1 Tank Design Considerations

To order a storage tank, the purchaser shall be well aware of the following essential considerations. As all these considerations have a bearing on design and thereby on the overall cost of the tank; clarity with regard to these requirements shall be of immense value to the owner of these capital-intensive assets.

Standards	
Establish suitable standards, codes, and specifications for the storage tank.	
Decide on an applicable design and construction codes, standards, and specifications to be given to the contractor such as STI Standard UL 142, API 12 series, API 650 Annex F, larger vertical cylindrical flat bottom tanks covered by API 650 without Annex F, API 620 and AWWA D-100 for water tanks of double curvature as per API 620, and so on.	
Site and Process Data	
1	Site-specific geotechnical data
2	Metrological data
3	Loading conditions for wind, snow, rain, or other loads
4	Physical properties for the range of liquids under consideration
5	Flow rate into and out
6	Any other special hazards associated with the stored medium
7	Other process and load data
8	Design life
9	Utility cost
Materials	
1	Corrosion and material compatibility
2	Establishment of cost factors and design life
3	Economic optimization of material (e.g., A S 73-70 plate costs 110% more to A 36 but provides 20% extra strength in allowable design stress)

Operational Data	
1	Tank capacity
2	Nature of service (including future changes expected like storage of another product, etc.)
3	Transfer rates
4	Access like ladders, platforms, means of sampling, use of floating suction lines, instrumentation requirements
5	Product purity
Liquid Properties	
1	Additional requirement above those already stated
2	Other problems associated with each fluid (e.g., total collapse of sulfur tanks due to a vacuum resulting from a blocked pressure relief valve [PRV]). Design in such cases often requires steam-traced vent lines and PRVs. Fully jacketed PRVs and vent lines are a solution to this issue.
3	Vapor pressure
4	Flammability and flash point (to decide on drive space and layout requirements)
Sizing Considerations	
1	Optimal size (most difficult task, usually taken as the sum of design capacity + maximum desired inventory + unusable volume at bottom)
2	Inventory (This means working capital. However, unexpected outages wipe out the advantages of minimizing design storage capacity and hence are considered as a trade-off based on the probability of such events.)
3	Comparison of raw material cost to risked cost of reordering, considering value discounts and bulk shipment advantages
4	Probability of accepting a full capacity shipment against a partial shipment
5	Cost of lost market share and goodwill due to insufficient capacity or unexpected outages or shutdowns
Venting	
Consideration for internal pressure surges as a result of explosions, deflagrations, exothermic reactions, decompositions, and similar events while providing safety relieving.	
Life Span	
Design life of the tank influences the design, in areas of corrosion allowance and use of linings and coatings and in providing a cathodic protection system within and below the tank.	

4.2 Design Aids Available

In earlier times, the design of storage tanks as per applicable codes was carried out manually according to methodology specified in code. Because of the advent of computers, software programs were developed by consultants and manufacturers on their own as in-house programs. Most of such custom-made software programs were developed in-house by consultants and manufacturers, whose authenticity was not professionally verified by any third party. However, such software was extensively used in the industry to carry out design based on the confidence level of the experts who developed it. One of the main disadvantages with these software programs was upgrading of the same in tune with revision of applicable codes. It is well known that codes undergo revisions periodically, that often call for periodic upgrading of software as well. The earlier scenario was that many of the small ventures that barged into the development of such software were not able to update their programs at this frequency on account of many reasons. However, with the stepping in of serious players from the software field in this kind of specific software, a lot of improvement has taken place over the past two decades. A few prominent software programs that have a proven record in the market and that carry out the design of storage tanks as per API requirements are listed in the following table.

Sl. No.	Software Trade Name	Coverage
1	E TANK 2000	API 650/API 620/API 653 and UL 142
2	INTERGRAPH TANK	API 650 and API 653
3	TRI* TANK	API 650
4	AMETank	API 650/API 620/UL 142 and AWWA

The advantage of these companies is that the sale of software is accompanied by updates to the software until the next revision of the code. In addition, these programs are capable of customization against wind conditions and other localized factors. This facility is often extended to end users of this software by suppliers, without which the software may produce unsound calculations for the applicable geographical zone where the proposed tank is to be constructed.

It is true that software can be bought at a price, but its effective, efficient, and economic use depends on experienced manpower available with end users and the kind of training and support imparted by software vendors. Please note that the availability of design software does not absolve the requirement of an experienced engineer having hands-on experience in carrying out the design of the storage tank manually to take care of complicated issues related to tank design. For experienced engineers with fundamental and sound basic knowledge of design, it would be very easy for them to feed the right data into a program based on purchase order and technical procurement specifications as the situation warrants. As accuracy of output is based on correctness and relevance of inputs, feeding in the right data as desired is of prime importance, and knowledge of basic design philosophy is of utmost importance in this regard.

While reviewing design calculations at the client's or consultant's end, most of the time, due to time constraints, a rerun of the program is not done. Instead a verification of input echo with requirements is spelled out in the technical procurement specification (TPS) for the tank, and the outputs for warnings, if any, are carried out at the client's or consultant's end.

As it stands, most of the design software programs provide a basic sketch of the tank along with the design calculations. Based on this and the scope drawing furnished along with the technical specifications, detailed fabrication drawings are developed by the manufacturer.

4.3 Basis for Designing

The basis for carrying out the design of storage tanks is as follows.

Data sheet and scope drawings provided along with technical specifications shall be the basic document as far as dimensions and other technical requirements are concerned. In addition, various conditions spelled out in the TPS by way of description and design criteria also shall be taken in to consideration, while carrying out design based on applicable codes. In most cases, special conditions shall be spelled out in the TPS, which would usually be above code requirements, as these are often user-specific requirements pertaining to the nature of the industry and fluid to be stored. This arises from the fact that codes and standards cannot formulate mandatory requirements for each and every type of tank in every industry. In fact, while formulating standards, a responsible technical committee addresses general minimum requirements within its ambit, and other finer details or requirements are left to the prerogative of the end user or owner of the facility.

Because of these reasons, a client or consultant working for a client shall have very clear information, especially regarding the additional requirements to be specified in the TPS or data sheet to be included in the inquiry. The easy way out in this regard is to overspecify requirements. While doing so, it shall be kept in mind that any overspecification that is not an essential requirement for proposed storage shall add to the cost of the tank, as nothing is available for free in this competitive world. This calls for judicious or rational framing of specifications based on specific and sound reasons.

4.4 Design Calculations

As of now, design calculations are produced by software (based on the data sheet, TPS, and scope drawing) and usually contain all salient details required to develop detailed fabrication drawings for all pressure retaining components and the load bearing structural members. These are submitted to clients or consultants for their review upon completion of the design and preparation of detailed fabrication drawings.

4.5 Drawings

Drawings are prepared by manufacturers according to their norms, and hence the total number of drawings required for construction of a storage tank varies with the manufacturer. However, the general practice followed by many manufacturers is to generate the following drawings as a minimum.

Serial No.	Description	Serial No.	Description
	For Fixed Roof Tanks	10	Manways: roof
1	General arrangement	11	Stairways and ladders
2	Annular plate and bottom plate	12	Handrails and platforms
3	Shell	13	Other attachments like pipe supports, etc.
4	Roof		*Additional for Internal FR Tanks*
5	Shell appurtenances	14	Details of floating deck
6	Roof appurtenances	15	Details of pontoons
7	Roof structures	16	Details of roof drain
8	Wing girders/stiffening rings	17	Details of seal
9	Manways/clean-out doors: shell	18	Leg supports (if required)

In the case of an internal floating roof tank, additional required drawings are listed as Serial No. 14 to 18 in this table. In the case of external floating roof tanks, roof structure and roof plate drawings will not be applicable. But such tanks can have other additions like rolling ladders, deck support, and so on, and many more drawings would be applicable as required.

Please note that required civil foundation and bund wall or dyke drawings are not considered under mechanical discipline and hence are not listed in this table.

The number of drawings required basically varies with the size of tanks and the intricacies involved in construction. For example, in the case of floating roof tanks, the total number of drawings required would be much like those listed in this table, as this list was developed for a simple fixed roof tank, wherein manufacturers have implemented standardization to some extent.

Construction drawings required for the erection of storage tanks include drawings from disciplines like civil, mechanical, electrical, and instrumentation disciplines. The rough scope of each discipline is shown in the following table.

Discipline	Subdiscipline	Scope
Civil		Foundation, dike wall, roads, drains, fences, other steel structures (in some organizations, the roof structure for the tank as well), pipe supports, and the like
Mechanical	Static equipment	Bottom plate, shell, roof plate sumps, nozzles, manways, wind girders, tank appurtenance (shell and roof), handrails, platforms, stairways, ladders, pipe supports on the tank, etc.
	Piping	Connected piping, internal piping, firefighting system
Electrical		Earthing, lighting, and other power connections and cabling including cathodic protection (CP) system
Instrumentation		All instrument connections and related cabling, including surveillance system

Since the scope of work varies from order to order, no generalization with regard to this table is possible. Moreover, the scope of other disciplines falls outside the ambit of this book and hence is not elaborated on further.

While preparing drawings for storage tanks, one should keep in mind that all essential requirements shall be spelled out in respective detail drawing. Furthermore, in this process, there shall not be any contradictions in requirements spelled out in different drawings. In this regard, it shall be made a practice to give the details or dimensions only at one location, and under no circumstances shall they be duplicated at a different location or in another drawing. During the process of preparation and review of drawings by various agencies involved in the construction of the tank, revisions will have to be incorporated for many reasons. If the same dimension is provided in more than once place in a drawing or in different drawings, it is a common phenomenon that revisions shall be incorporated only at one location, often leaving other places untouched and giving rise to a contradiction leading to the manufacture of components with wrong dimensions if different groups use drawings with conflicting dimensions. This simple recommendation is significant in tank construction, wherein a large number of dimensions and drawings are involved.

4.6 Approved for Construction Drawings and Documents

Drawings that are to be used for construction shall have an "approved for construction" (AFC) stamp on them. This is to be done by the design department of the fabricator after obtaining approval from the client or consultant for said documents. In case there is a holdup in finalizing some minute aspects of the tank, a drawing could be in the "approved as noted" stage with the client or consultant. In such cases, drawings may be issued for construction, with a hold indicated in the drawing for aspects pending decision. This will be necessary to meet stringent time schedules normally associated with such projects, though proceeding with "approved as noted" drawings is in violation of contract in most cases. Therefore, such advances, if any, made by contractor are credited under the contractor's responsibility. Most of the time, clients and consultants coerce contractors to proceed with construction with "approved as noted" drawings, duly considering the impact of "hold" indicated in the drawings.

4.7 Documents for Statutory and Client Approvals

Usually tanks are constructed as a part of a storage facility. In such circumstances, facilities warrant approval of statutory authorities of the province in which a tank farm is proposed. This may call for documents other than those for the tank, namely, the total facilities layout, along with drawings and documents pertaining to the safety of the farm as a total package, including fire-fighting systems. Since the purview of this book is the mechanical erection of storage tanks, documents pertaining to other works within the tank farm are not touched upon. In the case of an individual tank, manufacturers are expected to submit a copy of the design calculations along with all applicable drawings (design and drawings) for each tank to the client/consultant for approval prior to start of tank construction.

4.8 Design Change Note

In the course of the procurement, fabrication, and erection of a tank, minor design changes may be needed to accommodate site conditions or other unavoidable situations. Upon receiving intimation from a construction group about this problem, the design group has to take up this matter immediately with the client or consultant and arrive at an acceptable solution. In case the change to be made is of a minor nature, the same could be effected by a design change note (DCN) issued to all concerned. While issuing a design change note, the design group shall ensure that it reaches all original recipients of the affected drawing. In case the changes happen to be major, it is better to reissue a drawing with a new revision number duly indicating the zone of revision marked by clouding. Here also, receipt of the revised drawing by all original recipients shall be ensured. In case of design change notes, they shall be issued under continuous serial numbers, and all of them shall be compiled and incorporated in the respective as-built drawings to be made after completion of tank construction.

Upon receipt of a new revision of the document, the recipients are expected to destroy or return the superceded document to the issuer. Since this erases the history of revisions and the source, the originating department shall maintain records with regard to the revision history of each document it generates, with all salient details of its source. This information might be required at a later date to tackle some legal or money-related matters that may crop up during the execution phase of the contract.

4.9 As-Built Documents

As-built documents take care of all design changes (minor) incorporated after the issue of certified for fabrication drawings. The purpose of as-built documents is to provide a true reflection of the constructed tank including any substitution of materials (with concurrence) implemented during construction due to availability or other similar constraints. The release of revised drawings for every minor change required at the site (due to either engineering changes or construction needs) after issue of AFC drawings is a time-consuming activity. As mentioned earlier, in such cases, changes are incorporated through DCNs, which are controlled and documented and pass through necessary approval gates as required for the original design. The engineering or design team is responsible for incorporating all changes included in DCNs or otherwise (usually not recommended) and is expected to submit the same upon handing over the tank to the owner as as-built documents for future reference. Essentially, all engineering documents (drawings and design calculations) and any other construction-related documents considered relevant for future use shall be issued as an as-built document. Usually this is agreed upon prior to handing over the documents through a manufacturer's record book (MRB) index submitted to the client, which talks about the intended contents of the MRB. Most clients are specific in their requirements for MRBs with regard to contents, indexing, quality, and document mode like hard and soft copies and numbers required. Whether this is spelled out in the contract or not, each contractor is expected to provide minimum records of construction to the client as indicated in the standard. While this requirement is quite minimum and basic in nature, most clients ask for more documents that they consider relevant for their storage tank. However, for clarity in requirements, it is better to have discussions with the client at an early date in formulating the MRB documents along with progression of construction at the site. As of now, clients prefer searchable PDF electronic documents, in addition to one or two hard copies, which are considered absolutely essential.

5

Tank Foundation

5.1 Considerations for the Selection of Foundation

Annex B of API 650 provides important considerations for the design and construction of foundations for above ground storage tanks with a flat bottoms. However, because of a variety of factors such as surface, subsurface and climatic conditions, it is not possible to address all aspects pertaining to the design of different types of foundations in a comprehensive way. The allowable soil loading and type of subsurface construction are decided based on specific merits of individual cases.

For any tank site, the subsurface condition shall be known to estimate the soil-bearing capacity and probable settlement of the tank during the service life of the tank. An expert geotechnical engineer who is familiar with the history of the tank erection site can gather this information from soil boring, load tests, sampling, lab testing, and analysis.

A few salient conditions that require special attention as listed in Annex B of API 650 are as follows:

1. Tanks to be erected at hillsides where one part of the tank rests in an undisturbed area and the other part rests on filled-up land.
2. Tanks constructed on swampy land or filled-up land where there are layers of muck just below the surface.
3. Sites underlain by soils such as plastic clay or organic clay that may temporarily support an excessive load but may yield during long service at a rate greater than allowable rates.
4. Sites adjacent to water courses or deep excavations where the lateral stability of the soil is questionable.
5. Sites immediately adjacent to heavy structures where some of the load is distributed laterally to the soil below the tank.
6. Sites where tanks may be exposed to floodwater resulting in upliftment or displacement of the tank during service.

7. Sites in regions of high seismicity that may be susceptible to liquefaction.
8. Sites with thin layers of soft clay soil immediately below the bottom, which may cause lateral ground stability problems.

5.2 Types of Foundations

Since there is a wide variety of surfaces, subsurfaces, climatic conditions, and combinations thereof, it is not practical to establish design data to cover all situations. The allowable soil loading and exact type of subsurface construction required must therefore be decided for each individual case, after careful consideration, as elaborated above. As far as possible, the same rules, regulations, precautions, and considerations applied in the case of other foundations of comparable magnitude in the vicinity shall be adopted while designing the foundation for storage tanks.

Some of the usual types of foundations adopted for cylindrical tanks with uniformly supported flat bottoms are described next. However, a geotechnical study of the site is required to finally decide on the type of foundation and its design required at a particular site, taking into consideration the cost involved with the following options.

While it is difficult to classify all possible foundation types for storage tanks, listed next are a few general types that have proven over time to be cost-effective and efficient from a service point of view. These foundation types are listed in increasing order of costs.

5.2.1 Compact Soil Foundations

This type of foundation is generally preferred where soil quality and bearing capacity are good and is often called a *sand pad foundation*, laid directly on earth. Generally, existing topsoil (say from 100 mm to 150 mm) is removed first (depending on soil condition) and replaced with a sand or granular backfill. This is the cheapest foundation possible for any vertical storage tank with a flat bottom.

5.2.2 Crushed-Stone Ring Wall Foundations

This design incorporates a leak detection system. While it costs less than a concrete ring wall foundation, it has many of the advantages of the same. It provides uniform support of the tank bottom by distributing concentrated loads in a granular pattern. However, catastrophic failure of the foundation (and thereby the tank) is possible in the event of a leak from the bottom plate, which could wash away supporting matter.

5.2.3 Concrete Ring Wall Foundations

This type of foundation has a concrete ring wall around the tank periphery below the shell of the tank with a designed width, on which the shell rests over an annular or bottom plate. The rest of the foundation area shall be filled with sand and compacted. This type of foundation is used for tanks with a diameter of 10 m or more. For large-diameter tanks, this is the most cost-effective foundation, with many advantages such as a reduced probability of an edge settlement type of failure.

5.2.4 Slab Foundations

The concrete slab foundation is obviously the best foundation possible, but it is usually limited to tanks with diameters less than 10 m on account of its cost. Often the edge of the slab shall be thick enough to provide for anchorage as well (if required). A slab foundation is very versatile, but its high cost limits its use to small tanks. The slab provides a level and plane working surface that facilitates rapid field erection. Usually a leveling course of asphalt is provided on top of the slab to take care of any undulations on the surface caused during casting of the slab.

5.2.5 Pile-Supported Foundations

Pile-supported foundations are usually adopted in localities where the bearing capacity of the soil is very low. Tank locations in river deltas, backfilled land, land adjacent to bays, and so on could be considered as potential areas wherein this type of foundation may have to be adopted. Furthermore, this type of foundation is also considered necessary when high foundation uplift forces are encountered resulting from internal pressures or seismic loading.

5.3 Handing Over the Foundation

The civil engineering wing usually carries out the construction of the foundation. Upon satisfactory completion of work, followed by satisfactory inspection and testing, the foundation shall be released to the mechanical wing to proceed with erection of the tank. As there is a clear demarcation of work between the two disciplines, there shall be a proper handing-over document after successful completion of the foundation.

6

Sequence of Mechanical Works for Storage Tank Erection

6.1 Preliminary Works on Award of Contract

Design, fabrication, erection, inspection, and testing of storage tanks are highly specialized works that require a lot of experience and expertise. Because of this fact, storage tank erection work is usually awarded to reputed vendors with a proven track record. Since most of the above ground storage tanks are of large capacities, almost all manufacturing work takes place at the site. In this regard, the site manufacturing facility needs to be set up. Since this process takes some time, initial works related to design, preparation of drawings, inquiries for procurement, and so on are usually taken care of by the headquarters (HQ) of the contractor. Therefore, every construction contractor shall have a well-established work execution plan, wherein bifurcation of responsibility between HQ and the site team shall be clearly defined. A typical model that is usually followed by most tank construction contractors is provided next.

6.2 Responsibility Matrix (Head Office and Site)

Serial No.	Activity	Responsibility		Serial No.	Activity	Responsibility	
		HQ	Site			HQ	Site
1	Receipt of order	X		14	Design change requests (if any)		X
2	Data sheets (DS)	X		15	Design change notes (DCN)	X	
3	Design calculations	X		16	Data for as-built drawings and documents		X

(Continued)

Serial No.	Activity	Responsibility		Serial No.	Activity	Responsibility	
		HQ	Site			HQ	Site
4	Drawings	X		17*	Nondestructive testing (NDT) contract	X	X
5	Documents for statutory/client approval	X		18	NDT execution		X
6	Bill of materials (BOM)	X		19	Testing		X
7	Inquiry for plate, pipe, and other materials including consumables	X		20	Report of inspection and tests		X
8	Purchase of the above and delivery to the site	X		21*	Contract for surface preparation and painting	X	X
9	Receipt of raw materials at the site		X	22	Execution of surface preparation and painting		X
10	Storage of raw materials		X	23	Execution of internal lining if any		X
11	Manufacturing		X	24	Calibration of tank		X
12*	Welding procedure qualification and documentation	X	X	25	Compilation of site data		X
13	Welder qualification		X	26*	Preparation of manufacturer's record book	X	X

Note: *Optional between the two.

As mentioned earlier, the entire manufacturing work has to be carried out at the site. Therefore, setting up a temporary fabrication facility at the site is absolutely essential.

Activities at the site start with the arrival of the required machinery, raw materials, and consumables. This requires a lot of spade work at the head office of the manufacturer, namely, preparation of the design and drawings; preparation of other technical documents; placement of purchase orders for equipment (if any required), raw materials, consumables, etc., issue of various site subcontracts; and so on.

6.3 Organization Setup at the Site

6.3.1 Introduction

Construction companies with a proven track record in the erection of storage tanks by default should have established methodology in force to cover all activities right from submitting a quotation for the tank up to handing over the tank, usually within stringent contractual time frames. Effective and efficient progression of work requires well-defined systems within the organization, which is to be realized through a strong organizational structure within.

The erection of a storage tank is basically a site activity, and hence a temporary factory setup is required at the site. For any large contracting company, the erection of storage tanks shall form only a part of their activities, and the organization structure of the company shall be oriented toward those corporate goals. As a part of this corporate organization, a well-defined site organization structure is necessary to carry out all site activities according to the plans and procedures laid out for the purpose. Though the activities of the parent contracting company differ widely depending on their spectrum of operation, site-related activities in storage tank construction remain more or less the same, and hence site organization structure remains almost the same universally.

To complete all work related to tank construction with reasonably good quality (with an intention to avoid reworks), the contracting company must have a strong organization structure at the site with experienced and knowledgeable people manning all key positions. Organization at the site almost resembles that of any fabrication shop. However, for storage tank construction, the initial part of the work is carried out at the contractor's HQ, and all construction activities are to be completed using the site workshop (including forming of plates if required) and erection facilities. Therefore, a clear demarcation in the scope of the work proposed to be carried out by both groups shall be clearly laid out to avoid confusion during execution of the job with regard to the responsibility for each stage as described in Section 6.2. The organization structure recommended below is based on the following assumptions and presumptions:

1. It is presumed that the tank erection work is included under the scope of the contract.
2. Associated tank farm piping and other works are to be carried out by others.
3. Billing is done from the HQ (of the contractor) based on a duly certified progress report from the site.

4. Payment to large procurements shall be done directly by the HQ (of the contractor) based on material receipts at the site based on site inspection reports approved by the contractor and client.
5. Site purchases shall be limited to those items that are essentially required to proceed with day-to-day activities at the site.

6.3.2 Organization Chart

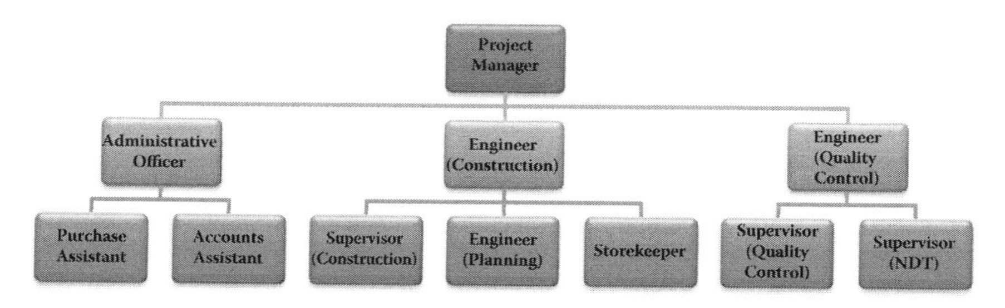

While the above organization gives an idea about various disciplines or work involved in site construction, their relative positions and numbers manning each discipline shall be decided based on the quantum of work involved. Furthermore, the site QC engineer shall report functionally to the corporate quality assurance/quality control (QA/QC) manager of the contractor (not indicated as it falls outside site activities) and may report to the project manager at the site only for administrative purposes.

6.4 Design, Drawings, and Documents Approval

As explained earlier, the design and approval of all related technical documents required to start work shall be the responsibility of the HQ group. Upon receipt of contract documents, engineering shall commence designing, followed by preparation of detailed fabrication drawings as listed in Section 6.5. The preliminary drawings thus prepared shall be reviewed by a group (joint review within the contractor) consisting of engineers from design, planning, QA/QC, and production departments with regard to all aspects related to construction. The comments of such review meetings shall be properly addressed by engineering (to make changes in drawings if required) as well as by other groups during various phases of erection. After incorporation of these comments, revision numbers A, B, — or 0, 1, 2, —, — shall be provided. The table indicating revisions history shall provide a brief description of the salient revision and reasons for the same at a designated location of the drawing to provide an overview of revisions implemented. Areas or regions where revisions were implemented shall be highlighted by clouding (for drawings) or by a straight line in the margin (for text documents) to help reviewers locate revision zones and spot changes easily. Earlier versions of documents shall be marked as either "superceded" or "obsolete," and these documents shall be withdrawn from respective end users of these documents. However, all end users shall have a system of maintaining superseded documents for verification of changes made at a later date for resolving commercial or other related legal matters, and hence these documents shall be available only to those authorized to do so, preferably the concerned section head.

At this stage, the prepared documents are ready for submission for approval from various agencies involved such as the client, consultant statutory authorities, and so on as applicable. These documents are treated as deliverables under contract, and a proper tracking system shall be in position for proper monitoring and control of the approval process by various agencies concerned. For easy tracking and control of revisions of documents, it is also advisable to develop a proper document numbering system for all documents being generated for the work undertaken by the contractor. The document numbering system (including that for drawings) for storage tank construction shall resemble those adopted in other business units under the corporate umbrella of which the tank construction forms a part.

It is preferable to send documents simultaneously to all reviewing agencies for their approval. However, few back-and-forth transmissions would be required to reach "approved for construction" (AFC) status by all agencies concerned. When submitted for approval, comments on drawings are usually given as a markup in red color, whereas that for text documents is provided as a comment sheet or as markup at the option of the reviewer.

If the comments offered are within contractual requirements, they are incorporated in the next revision of the document and submitted for final approval of those concerned. Whereas if the comments fall outside the contract, the contractor is eligible to claim extra on account of the same and hence to be notified to the client through a "compliance sheet" enclosed with the revised document. As mentioned earlier, incorporation of comments (with or without reservations) shall be the responsibility of the origination section of that particular document. Therefore, the respective departments shall be responsible for maintaining a history of revisions implemented as well. Furthermore, they shall also ensure that revised documents have reached all end users, immediately after release of the new revision.

6.5 List of Documents

Sl. No.	Description of Document	Approval				
		A	B	C	D	AB
1	*Engineering*					
1.1	Design Calculations	X	X	X		X
1.2	Drawings					
1.2.1	General Arrangement Drawing	X	X	X		X
1.2.2	Bottom Plate Layout and Weld Details	X	X	X		X
1.2.3	Bottom Plate Cutting Plan	X	X*	X*		X
1.2.4	Shell Development and Shell Weld Details	X	X	X		X
1.2.5	Shell Plate Cutting Plan	X	X*	X*		X
1.2.6	Roof Plate Layout and Weld Details	X	X	X		X
1.2.7	Roof Plate Cutting Plan	X	X*	X*		X
1.2.8	Dome Roof Details	X	X	X		X
1.2.9	Roof Structure Details	X	X	X		X
1.2.10	Draw Off Sump Details	X	X	X		X
1.2.11	Nozzle and Manway Schedule	X	X	X		X
1.2.12	Special Notes	X	X	X		X
1.2.13	Nozzle Weld Details (Shell and Roof)	X	X	X		X
1.2.14	Clean-out Door Details	X	X	X		X
1.2.15	Manway Weld Details (Shell and Roof)	X	X	X		X
1.2.16	Manway Davit Details	X	X	X		X
1.2.17	Wind Girder Details	X	X	X		X
1.2.18	Stairway Details	X	X	X		X
1.2.19	Ladder and Platform Details	X	X	X		X
1.2.20	Hand Rail Details	X	X	X		X
1.2.21	Details of Inlet Distributor Pipes	X	X	X		X
1.2.22	Details of Suction Header	X	X	X		X
1.2.23	Details of Other Internal Attachment	X	X	X		X
1.2.24	Details of Other External Attachments	X	X	X		X
1.2.25	Details of Piping Clips	X	X	X		X
1.2.26	Details of Deflectors	X	X	X		X
1.2.27	Details of Foam System	X	X	X		X
1.2.28	Anchor Chair Details	X	X	X		X
1.2.29	Details of Foam Cooling System with Foam Makers and Orifice Plate	X	X	X		X
1.2.30	Details of Fixed Water Spray Cooling System with Deflector and Supports	X	X	X		X
1.3	Floating Roof Details	X	X	X		X
1.3.1	Details of Rolling Ladder, Track, Earthing, etc.	X	X	X		X
1.3.2	Details of Roof Supporting Structure, Buoys with BOM	X	X	X		X
1.3.3	Details of Primary and Secondary Seal and Details of Earthing Shunt	X	X	X		X

Sl. No.	Description of Document	Approval A	B	C	D	AB
1.3.4	Details of FR Appurtenance, Roof Stopper, and Leg Supports	X	X	X		X
1.3.5	Orientation of FR Appurtenances, Leg Support, Rolling Ladder, etc.	X	X	X		X
1.3.6	Details of Primary Roof Drains and Drain Sumps	X	X	X		X
1.3.7	Details of Still Wells, Gauge Hatch	X	X	X		X
1.3.8	Mounting Details of Rim Vents	X	X	X		X
1.3.9	Details of Emergency Roof Drains	X	X	X		X
1.3.10	Details Anti-Rotation Device	X	X	X		X
2	*QA/QC Documents*					
2.1	Construction Quality Plan	X	X	X		X
2.2	WPS/PQR/WQT Records	X	X	X		X
2.3	Tank Erection Procedure	X	X	X		X
2.4	Welding Consumables Control Procedure	X	X	X		X
2.5	Hydrostatic Test Procedure for Completed Tank	X	X	X		X
2.6	Pneumatic Test Procedure for Roof Welds	X	X	X		X
2.7	Pneumatic Test Procedure for Reinforcement Pads	X	X	X		X
2.8	Procedure for Calibration of Welding Equipment	X	X	X		X
2.9	Procedure for Calibration of Electrode Oven	X	X	X		X
2.10	Hardness Test Procedure	X	X	X		X
2.11	PWHT Procedure	X	X	X		X
2.12	Vacuum Box Test Procedure	X	X	X		X
2.13	Settlement Measurement Procedure	X	X	X		X
2.14	Inspection and Test Plan (Mechanical Works)	X	X	X		X
2.15	Surface Preparation and Coating Procedure for Internal	X	X	X		X
2.16	Surface Preparation and Coating Procedure for External	X	X	X		X
2.17	Inspection and Test Plan (Surface Preparation and Painting/Coating)	X	X	X		X
	NDT Procedures					
2.18	Liquid Penetrant Test Procedure	X	X	X		X
2.19	Magnetic Particle Test Procedure	X	X	X		X
2.20	Ultrasonic Test Procedure	X	X	X		X
2.21	Radiographic Test Procedure	X	X	X		X
2.22	Visual Examination Procedure	X	X	X		X
3	*Construction Records*					
3.1	Material Summary Report and Material Map				X	X
3.2	Material Test Certificate (All Pressure Parts and Internals)				X	X
3.3	Bottom Plate Weld Map				X	X
3.4	Vacuum Box Test Reports				X	X
3.5	Weld Map				X	X

(Continued)

Sl. No.	Description of Document	Approval				
		A	B	C	D	AB
3.6	Weld Inspection Summary				X	X
3.7	Weld Map and NDT Summary				X	X
3.8	NDT Reports				X	X
3.9	Pad Air Test Reports				X	X
3.10	BOM Check Report				X	X
3.11	Hydrostatic Test Report				X	X
3.12	Settlement Report and Cosine Curve				X	X
3.13	Signed Off ITP for Erection Work				X	X
3.14	Surface Preparation and Painting/Lining Reports				X	X
3.15	Adhesion/Other Coating Inspection Test Reports				X	X
3.16	Signed Off ITP for Surface Preparation, Painting/ Lining				X	X
3.17	Tank Calibration Reports				X	X
3.18	Handing Over Report				X	X

Note: A = consultant, B = client, C = statutory authority, D = site team, AB = as-built. (1) All documents listed under Clauses 1.2, 1.3, and 2 may not be applicable in all cases, as the list provided is applicable for both floating and fixed roof tanks. (2) The documents listed under Construction records need to be compiled along with work progress. (3) The documents indicated mainly refer to the lead document only. Each such document shall have many attachments as well. For example, the pneumatic test report shall have a valid calibration certificate of the pressure gauge used for testing.

* Desirable

It shall be noted that comments offered by different agencies shall be strictly within agreed terms of the contract or in accordance with statutory regulations (which shall be a part of the contract), otherwise there is every chance that the contractor could swing back with an extra claim. Depending on the quality of Revision "A" documents in respect to compliance to code, the technical procurement specification (TPS), and statutory requirements, it is possible that a document could reach AFC status in one transmittal. In reality, the first submission of documents is usually done just to meet the contractual obligation of submitting deliverables, and because of this, a lot of omission might be there that may call for a few more submissions for the documents to achieve AFC status. Strictly speaking, most of the purchase orders placed shall insist that contractors start work only after obtaining approval from the client or consultant. Strict compliance to this requirement may lead to time overruns, and hence in most cases, contractors are permitted, at their risk and cost, to proceed with procurement (especially of long lead items) and other similar actions even before approval of necessary engineering documents.

6.6 Preparation of Bill of Materials with Specifications

After completion of the design, detailed fabrication drawings as listed in Section 6.5, are prepared. Each drawing is expected to carry a bill of materials (BOM) for components covered by that drawing. For example, a manway detailed drawing shall contain a BOM requirement for the manway, including a davit and so on as required. Similarly, a shell development drawing shall contain the BOM for all shell courses, whereas a general arrangement (GA) drawing is not expected to carry a BOM. However, it shall contain all salient common requirements including a nozzle schedule and list of accompanying documents, reference drawings, documents, and so on, for correlation purposes. Salient information normally included in a GA drawing for a storage tank designed as per API 650 is listed in the following table.

Details Shown in General Arrangement Drawing

Sl. No.	Description of Document	Sl. No.	Description of Document
1	Design code and other salient inputs	5	Nozzle/manway schedule
2	Special notes specific to the tank applicable in entirety	6	Basic material of construction like that of an annular plate, bottom plate, shell roof, etc.
3	Reference documents	7	Basic design considerations
4	List of detailed drawings	8	NDT and other salient tests required

6.7 Inquiry and Purchase of Raw Materials

Based on the BOM developed for each tank, a consolidated quantity of materials (in case the order is for more than one tank or including associated works such as piping, etc.) is prepared with details such as material specification, quantity, thickness/schedule, and so on, with special requirements if any against each material required with some excess. Many times special requirements for components shall be those specified in data sheets or specification (TPS), which in most of cases falls under supplementary requirements in respective materials specifications. By providing simple material specification in a purchase order (like ASTM A 106 Gr B), the vendor is liable to comply only with the basic minimum requirements of A 106 Gr B, which means none of the supplementary requirements need be complied by the vendor. Therefore, care shall be taken while ordering materials, with all requirements spelled out unambiguously in the order.

Inquiry is usually sent to a few renowned manufacturers (for bulk materials) and to reputed traders for small quantities of materials required for nozzle pipes and fittings and so on as required. Whether the material is procured from mills or traders, it shall meet all chemical or physical requirements spelled out in applicable codes and as restricted by the purchase order. Furthermore, materials procured from traders shall have positive traceability with regard to hard stamping on materials and endorsement in related material test certificates (MTCs) for the materials.

6.8 Materials Procured from Stockists or Traders

The following aspects shall be considered absolutely essential in the case of small quantities of materials such as pipes, fittings, flanges, fasteners, and so on to be procured from stockists for construction of storage tanks. Compared to plate materials required for storage tanks, requirements for piping elements shall usually be very small, and hence procuring them from manufacturers would be often impractical. Therefore, the option available shall be to approach stockists who can supply these items with proper material certification. The following minimum requirements shall be complied to while doing so, in order to ensure that the material procured is genuine and traceable:

- Since stockists will have bulk materials in their stock, they may not provide the original MTC for small quantities of materials procured from them. In that case, it shall be ensured by the buyer that the original MTC is available with the stockist.
- The third-party inspector (TPI) or client representative assigned to stockists yards to inspect materials prior to dispatch shall be instructed to make such an endorsement either in copy of the MTC provided or in the inspection release note (IRN) issued by him or her after the inspection.
- Materials shall be traceable to the MTC through proper stamping or stenciling on raw materials such as pipes, fittings, and flanges as required in applicable specifications and manufacturing practices and endorsed accordingly.
- The TPI shall ensure that the origin of materials is from renowned sources with proven records.
- The TPI shall ensure that additional testing and certification requirements (as specified in the TPS) are also certified in the MTC. If not, this shall be reported as a discrepancy, and the buyer's confirmation shall be sought prior to release of materials.
- In case it is agreed by the buyer that the discrepancies can be covered by additional testing at the construction site, discrepancies shall be reported in the IRN, with a reference to the buyer's clearance to release with the reported discrepancy.

6.9 Purchase or Transfer of Construction Equipment to the Site

Most tank construction contractors shall have a stock of machinery required for site erection. However, in some instances, they may need to procure new equipment depending on specific requirements for the project. Since some pieces of such equipment are long lead items with high capital investment, a fairly accurate assessment of equipment requirement is absolutely essential (with a clear vision for the future) for smooth progress in construction. Timely availability of all equipment is required at the site too, as required numbers shall be a critical factor in timely completion of the project. The following table shows equipment that is usually required at a site for construction.

6.10 List of Equipment Required at the Site

Sl. No.	Equipment Description	Sl. No.	Equipment Description
1	Plate cutting machines (auto and manual)	18	Gouging machines
2	Grinding machines	19	Drilling machines
3	Plate bending machine	20	Hydraulic press to form gratings
4	Lifting tools and tackles	21	Radiographic equipment (gamma or X-ray)
5	Welding machines (transformers, rectifies, MG sets, etc.)	22	Ultrasonic equipment with probes and accessories
6	Water pressure pump	23	Magnetic particle test equipment
7	Transport equipment (trucks and pickups)	24	Dye penetrant test kits
8	Hydraulic jacks and power pack with controls	25	Magnifier and portable lights
9	Pulley blocks	26	Vacuum box testing equipment
10	Hoists	27	Water pump
11	Hydraulic wheel mounted cranes 50 and 15 T	28	Air compressor
12	Pipe bending machine	29	Mother and portable ovens
13	Measuring tapes 100, 30, 15, 5, and 3 m	30	Vernier calipers
14	Piano wire, plumb wire	31	Film viewer
15	Aluminum ladders	32	12 V hand lamp sets
16	Exhaust fans	33	Man coolers
17	Compressor for pneumatic tools	34	Water coolers

Note: NDT equipment listed as Sl. No. 21 to 23 can be outsourced through a subcontract.

6.11 Work Contracts

In case site erection work is directly taken up by the contractor, at a minimum the following subcontracts would be required to complete construction as per specification. It is presumed that such contractors shall be available in the vicinity of the site where the tank is to be constructed, and if so, it would be cheaper to use them rather than use a contractor's own team, to be mobilized from elsewhere, most likely from HQ.

The usual practice is to have the following subcontracts organized at the site, from cost, convenience, and time frame considerations.

6.11.1 Subcontract for Nondestructive Testing (NDT)

All NDT work required to be carried out on tanks, such as radiographic testing (RT), ultrasonic testing (UT), magnetic particle testing (MPT), and dye penetrant testing (DPT), is usually covered in this contract. However, some contractors might wish to carry out a low-end NDT such as MPT and DPT with their own staff. This is because these tests may often be required at short notice during erection, wherein availability of technicians from an NDT subcontractor might not be possible within short notice. In such instances, it shall be the duty of the main contractor to provide NDT technicians with the requisite qualifications to carry out, interpret, and document these NDT as required in applicable code and specifications.

6.11.2 Subcontract for Scaffolding

Yet another usual subcontract is for scaffolding. Many tank manufacturers have their own scaffolding services as well, whereas some prefer to go for subcontracts depending on logistics and cost.

6.11.3 Subcontract for Surface Preparation, Internal Lining, and External Painting

All tanks require surface preparation and external painting as per the client's specifications. Some tanks also may require some special internal lining as well. Usually a package consisting of surface preparation and application painting or lining is subcontracted to an expert contractor who is specialized in this field. These applicators, especially those needed for lining application, need to be trained and qualified as per manufacturer's recommendations, which would also be a part of the subcontract.

6.11.4 Calibration and Certification of Storage Tanks

Calibration and certification involving statutory authorities is usually yet another specialized job requiring expertise and hence off-loaded as a subcontract package for convenience.

At the discretion of the management of the contractor, these subcontracts can be arranged directly from HQ or at the site. If this is done at the site, the project manager shall be responsible for all activities related to subcontracting, as well as the execution of the same.

6.12 Local Contracts and Purchases at the Site

Any other contract that is of not much significance to construction such as scrap removal and stacking of plates; housekeeping; trucking; maintenance support for equipment and consumables such as oxygen, acetylene, argon, and so on; and other consumables for the day-to-day activities at the site also shall be essentially initiated and controlled at the site so as not to hamper production. The project or construction manager at the site shall be responsible for such local contracts and purchases.

7

Tank Erection

7.1 Storage Tank Erection

The references made here are mainly applicable to normal cone roof storage tanks, as they are simple in construction and the most common in the process industry where liquid storage is essential. As explained earlier, the location of the liquid storage system for a plant is decided based on many economic considerations and is not included within this book. Furthermore, it is presumed that the foundation is completed and ready for the laying of the bottom plate, as well as all subsequent activities related to the mechanical erection of the tank. While handing over the foundation for the mechanical erection of the tank, it shall be the responsibility of the civil engineer in charge of the foundation and other civil works to carry out all tests specified to ensure that the foundation has achieved requisite strength as specified and considered for the design of the foundation and the tank.

As mentioned earlier, the most commonly used type of foundation is the sand pad type with bitumen topping to obtain the required slope and surface. The sand pad foundation with a concrete ring wall foundation (with bitumen topping) is also extensively used in the industry. In some specific cases, a concrete pad foundation is also used, again with a bitumen layer of 50 mm applied over the concreted surface to obtain a uniform surface for proper laying of the bottom plate. The decision to go for a particular type of foundation is based on economic considerations, as briefly described in Chapter 5. Irrespective of the type of foundation used, the mechanical erection methodology of the tank remains unaltered and is described in the following sections.

7.2 Inspection of Raw Materials

All materials used in the manufacture of a storage tank shall be traceable to its documentation, especially that for all pressure retaining parts. These materials are to be identified against each of the items indicated in applicable drawings. It is presumed that all specific criteria required in code or client specifications are taken care of while preparing detailed drawings. If this is done, it is easy for the rest of the departments to simply follow the drawings in letter and spirit to meet all client requirements. This assumption should be true in all respects so that at the time of receipt of materials, there are no ambiguities regarding documents to be presented by the vendor for verification as required in the inspection and test plan (ITP). For inspection of any raw materials arriving at the site warehouse, purchase orders and technical procurement specifications shall form the basis for carrying out inspection.

The format enclosed as Table 7.1A may be used to inform the concerned departments about the receipt of materials at the site. On the basis of this notification, the QC inspector or engineer shall carry out inspection of raw materials and report his or her findings back to all concerned through a

TABLE 7.1A

Intimation of Material Receipt (IMR)

Purchase order no.	Date
Vendor	Date of receipt

Details of items received

Sl. No.	**Tag No.**	**Description**	**Qty**	**Remarks**
1				
2				
3				
4				
5				
6				
7				

Location	
Status (MTC)	Received/not received

Warehouse in charge	Name		Signature		Date	

Routing	Engineer (QC)
Cc	Inspector(QC)
Cc	Purchase assistant
Cc	Engineer (planning)
Cc	Engineer (construction)
Cc	O/c 1 Master File
Cc	O/c 2 PO file

TABLE 7.1B

Material Receipt Inspection Report (MRIR)

Purchase order no.		Date	
Vendor		Date of receipt	
IMR reference		IMR date	

Details of items inspected

Sl. No.	Tag No.	Description	Qty	Remarks
1				Accepted/Hold/Rejected
2				Accepted/Hold/Rejected
3				Accepted/Hold/Rejected
4				Accepted/Hold/Rejected
5				Accepted/Hold/Rejected
6				Accepted/Hold/Rejected
7				Accepted/Hold/Rejected

	Clarifications needed for removal of hold
1	
2	
3	
	Reasons for rejection
1	
2	
3	

Inspector (QC)	Name		Signature		Date	

Routing	Warehouse in charge
Cc	Engineer (QC)
Cc	Inspector(QC)
Cc	Purchase assistant
Cc	Engineer (planning)
Cc	Engineer (construction)
Cc	O/c 1 Master file
Cc	O/c 2 PO file

Material Receipt Inspection Report Table 7.1B. Upon satisfactory completion of the inspection at the site, it is considered that the materials comply with all requirements spelled out in applicable material specification, as well as supplementary requirements called for through client specifications.

The following activities shall be applicable to all raw materials to be used in construction of the storage tank, namely, plates, pipes, fittings, flanges, fasteners, and structural steel members that are directly intended to bear loads or be in contact with the medium stored:

- Physical verification of identification hard stamped/stenciled on plates, pipes, fittings, flanges, and so on, such as specification, heat number, stamping details of inspection agency, and so on
- Cross verification of the above data with respect to details given in respective mill test certificates (MTCs) and original stamping
- Verification of test reports with respect to code and purchase order requirements
- Physical verification of the condition of the material for any transit damages and aging in the case of materials procured from traders

When all these aspects are in order, the material receipt intimation note can be closed by issuing a material receiving inspection report with the inspectors' remarks on it and also by attaching a green tag on it Figure 7.1A. The warehouse department shall take only such materials in stock, and only those in stock shall be released for subsequent construction activities upon request by the production team.

As mentioned, upon receipt of the materials at the site, the stores section shall inform the QC department about the receipt through the intimation note as per format Table 7.1A. In the meantime, the procurement section is expected to forward all relevant documents (MTCs and other certificates if any) pertaining to materials to QC to facilitate inspection, with reference to the purchase order and intimation report (issued by the warehouse section), with a copy provided to planning for advance actions at their end. Until inspection activity is completed, raw materials are kept in the "incoming material inspection bay" of the warehouse (for all materials other than plates and pipes, which are usually stored separately in an open yard allocated for bulk material storage).

Items that have undergone inspection shall essentially have a green, yellow, or red sticker or a tag with the initials of the inspection engineer (and with reference to the receiving inspection report) who attended the inspection, with the date of inspection as shown in formats Figure 7.1(a), (b), and (c). The green sticker or tag is for "accepted" items, yellow is for "under hold" items, and red is for "rejected" items.

Items kept under hold are mostly on account of some missing information or document to fulfill PO requirements. All such deficiencies shall be intimated to the concerned department (planning, procurement, design) in detail through an inspection report as per format Table 7.1B. The concerned sections mentioned previously shall be responsible to remedy the shortfall appropriately as reported in the inspection report and shall communicate with the vendor or subcontractor with clear actions to remedy

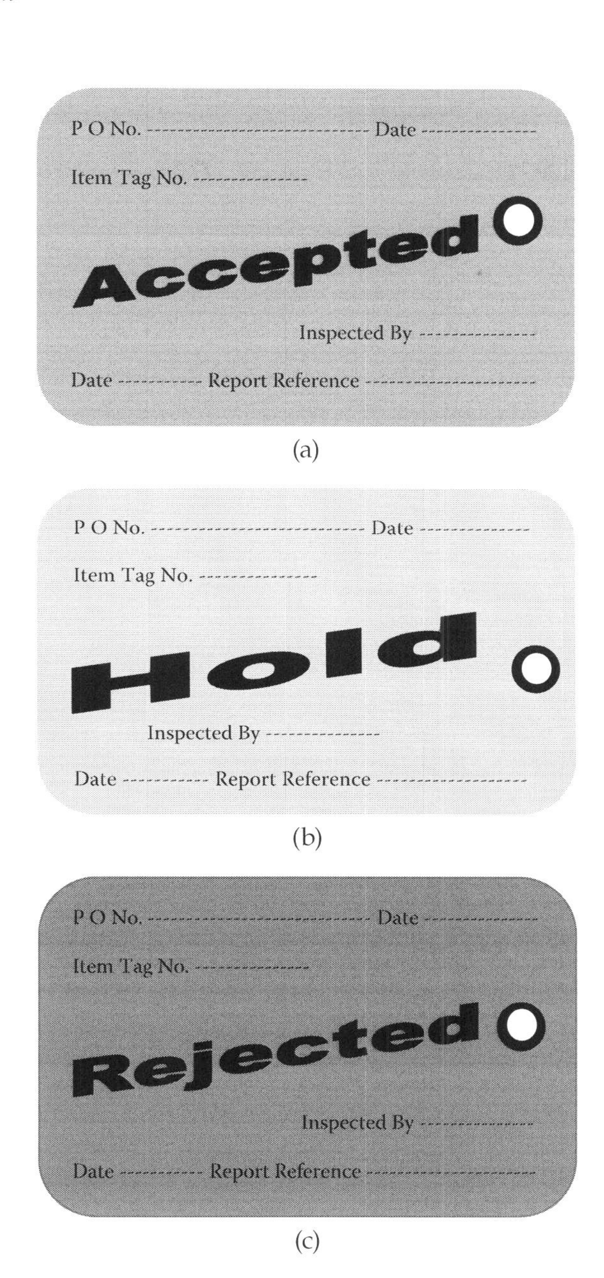

FIGURE 7.1
(See color insert.) Inspection status tag/sticker.

the shortfall. Items shall remain under hold in an area earmarked for the purpose (quarantine area) till the matter is resolved. Items under hold shall not be taken in stock until the hold is removed and the item moves to the "accepted" category with a green tag. This system could ensure that all accepted materials have a green sticker or tag, which would imply the following:

- The materials are suitable for use as per the purchase order and technical specifications.
- The MTCs furnished are in line with the requirements.
- The materials supplied are traceable to the certificates furnished.

Items that do not meet PO or code requirements shall be rejected and hence shall have a red sticker or tag. The reasons for rejection shall be elaborated in the inspection report and need to be taken up with the vendor by the procurement section for replacement. At the same time, the warehouse section shall take steps to return rejected materials back to the vendor at the vendor's risk and cost.

As mentioned, the predominant raw materials for a storage tank are plates, pipes, flanges, and fittings. The recommended inspections for specific raw material types are elaborated to ensure quality and to avert untoward revelations at a later date.

7.2.1 Plates and Pipes

Measure possible physical dimensions, especially thickness. Correlate identification found on these items with that provided on the certificate furnished. In the event all of the above is in order, the certificate is verified against the code and PO requirements. If this also is in order, items can be accepted.

7.2.2 Pipe Fittings and Flanges

Record all possible dimensions and verify them against those given in applicable standards. Cross-check identification found on items to that given in the certificate. Verify the certificate against PO and code requirements. In case of large quantities, 100% inspection may not be possible or be rather a waste of time, provided materials are procured from proven reliable sources that have a well-established QA/QC organization and structure. Therefore, a random sample inspection can also be resorted to, for which a 10% random inspection at receipt is adopted almost universally. In case deviations are observed in this random inspection, a progressive inspection shall be imposed, resulting in 100% inspection eventually, at which stage the reliability and dependability of that vendor is questionable. When randomly selected samples from a lot meet all requirements, the entire lot is accepted. Similarly when a progressive examination results in rejection, the entire lot shall be rejected. The mediocre policy of accepting the tolerable components from a lot subsequent to high rejection is not considered as a reliable methodology in random inspection.

7.2.3 Fasteners and Gaskets

Select random samples (say at 10%) and subject them to a dimensional check. Correlate certificates to identification given to the lot. Verify certificates for compliance to PO and code requirements. Once samples are found satisfactory, the lot represented by the sample shall be accepted.

7.2.4 Consumable Such as Electrodes

The containers for consumables such as electrodes are not to be opened for inspection. In such instances, inspection shall be limited to verification of batch numbers provided on containers, followed by cross matching them with certifications provided. If both match, items can be accepted. Furthermore, the condition of containers carrying electrodes upon receipt also shall be designated as a matter of concern, as it affects the quality of the weld on prolonged storage.

7.3 Identification and Traceability of Raw Materials

It is essential to maintain traceability records for all components of storage tanks, especially those for pressure retaining parts and those in contact with the stored medium. This needs to be carried out progressively and shall be available for each pressure retaining part identified in the relevant drawings. Therefore, the traceability methodology shall be evolved within the organization to cover each and every part number identified in the drawings, with a bill of materials (BOM) in each drawing:

- Obtain mill test certificates (MTCs) from the planning or procurement department (concerned team) for all materials issued for production.
- In the case of plates, if used in full size, original stamping shall be retained on respective components as such. If not, stamping as seen in the original material shall be transferred to parts marked on each plate, preferably accompanied by the personal stamp of the inspector responsible for the transfer of identification.

- Identify certificates for plates used for components and compile them in the MTC file pertaining to the respective tank with proper identification of the specific part for which the plate is used.

This methodology shall be maintained for all other raw materials, where a few parts are taken from one piece of raw material. Whereas for materials such as fittings, flanges, and so on, one certificate will cover many items, and in such cases, a list of part numbers for which this MTC is applicable shall be maintained as a table for easy understanding or by any other methodology deemed fit for positive and easy traceability. Furthermore, some clients may require hard stamping on individual components on the external surface. In such cases, please note the following:

- Hard stamping shall be limited to carbon steel materials like plates, pipes, fittings, and so on.
- Always ensure that these stampings are carried out on the external surface, wherein no contact with service fluid is expected.
- For stainless steel materials, chemical etching shall be resorted to for transfer of identification.
- For all types of flanges, hard stamping is possible on the flange edge, which in no way hampers its service life and is hence recommended.
- When a strong QC system is in force at the site, and if agreed by the client, consultant, and manufacturer, the transfer of identification using paint markers can also be considered. All transferred identification shall also contain the personal initials of the QC inspector who witnessed the transfer of identification.
- If permitted, identification transfer using paint markers shall be clear and legible, and a designated color shall be used for this purpose to distinguish it from other possible markings that may come later on the component.
- When paint markers are used on stainless steel components, it shall be ensured that paint used is free from chlorides and suitable for marking on stainless steel.

7.4 Marking of Plates

Cutting plans for all plates (bottom, shell, and roof) are prepared by the design team with a view to optimize utilization. Ordering the right size of plates (based on logistics and other constraints) needs to be given due consideration, based on the final dimensions of the tank agreed upon by the purchaser and contractor. The usual plate size available in the market is around 12,000 mm × 3,000 mm depending on mill limitations. It is always better to get the largest size possible for tank work so that weld joint length can be reduced significantly, thereby reducing the cost of construction. While considering plate width for ordering, consider the plate bending machine capacity (especially maximum width) of models available in the market.

7.4.1 Marking of Annular, Sketch, Bottom, and Roof Plates

Marking annular and sketch plates shall be carried out as specified in the cutting plan. In the case of full plates used for bottom plates, squaring of the plate is not really required, provided plate edges are free from detrimental defects, as the type of joining between annular, sketch, and bottom plates is through lap welding.

In the case when a few smaller sketch plates are marked from a full plate, the material identification stamping found on the plate shall be transferred to all components that are to be made from each plate. As a standard practice, almost all renowned manufacturers provide two stampings on each plate. In case such stampings fall on a component marked, such segments do not require further identification transfer. To have proper control over identification transfer, it is preferable to place the personal stamp of the inspector concerned as well, along with the transferred identification.

7.4.2 Marking of Shell Plates

No special marking is required for any of the full shell plates, required to make one circumference. However, squareness or perpendicularity of plates needs to be checked. Squareness of shell plates can be checked by measuring two diagonals of the plate. The methodology to measure diagonals without error is provided next.

Consider a plate of 12,000 mm × 3,000 mm with a sheared edge. In all probability, the edge of the plate can have a slight taper on account of the shearing carried out at the mill for sizing. Therefore, if the measurement of the diagonal is made from the sheared edge of the plate, the chance of error is more, and hence it is recommended to measure the diagonal at points 100 mm away from all sides, as shown in Figure 7.2.

If the difference between the two measured diagonals of the plate shown as D1 and D2 in Figure 7.2 falls within ±2 mm, the plate can be considered square for all practical purposes.

Marking annular, sketch, and odd-shaped roof plates shall be carried out as specified in the cutting plan. For all components, the material identification stamping found on the original plate shall be transferred to all components that are to be made from the plate.

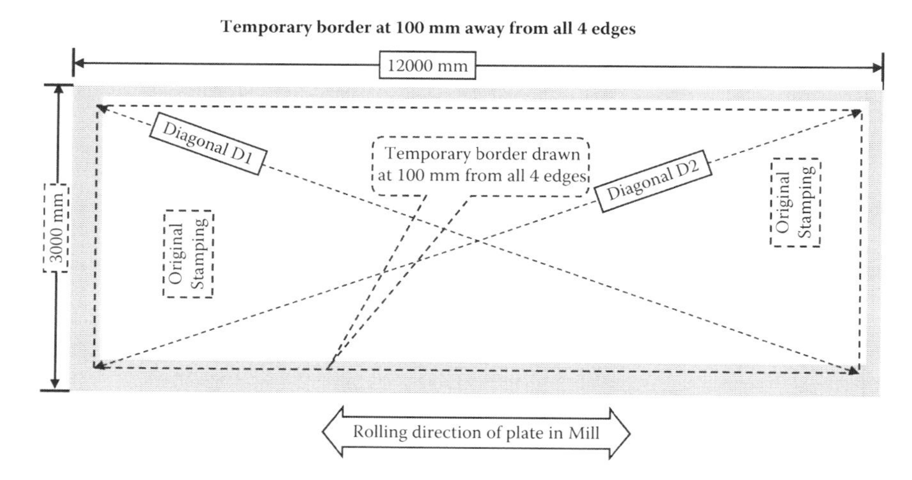

FIGURE 7.2
Method to verify squareness of plates.

7.5 Cutting of Plates and Other Pressure and Load Retaining Parts

For the majority of storage tanks, the material of construction would be carbon steel of varying grades depending on the service conditions. However, storage tanks made of exotic materials are also required in the chemical industry, especially to store chemicals that are highly corrosive to carbon steel. For carbon steel tanks, plate cutting to the required shape and dimension shall be by oxyacetylene cutting, especially for irregular shapes. Manual, simple, portable semiautomatic and automatic with computer numerically controlled (CNC) machines are available in the market for this purpose. These machines are capable of producing not only a square edge but also "V" (single and double "V") bevels as in Figure 7.3(a) and 7.3(b), at required angles, whereas it cannot produce double bevels as in Figure 7.3(c). For all types of materials (carbon steel and stainless steel), shearing plates is an option, in case of square and straight cut lines. For cutting contours, shearing is not possible, and other means have to be used. When exotic materials are involved, plasma cutting is yet another option, based on the thickness of the plate to be cut.

While much preparation is not required for plates cut by shearing, because of oxidized material present on the cut edge, some amount of preparation shall be required in case of oxyacetylene and plasma cut plate edges.

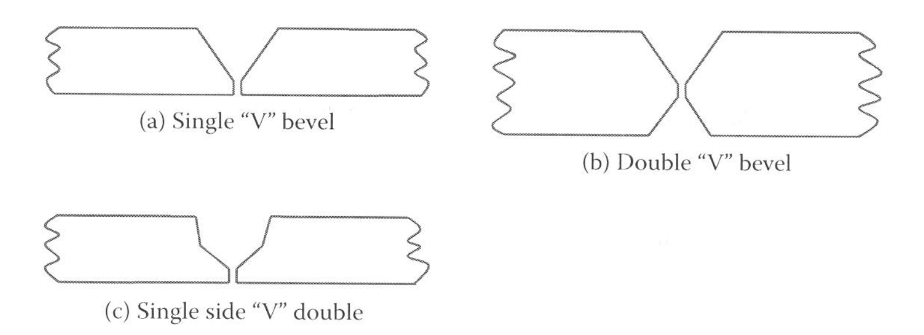

FIGURE 7.3
Typical bevel edge preparations.

7.6 Inspection of Plates after Cutting and Edge Preparation

As mentioned, the most commonly used process for cutting and edge preparation of carbon steel plates is oxyacetylene cutting, and that for stainless steel plates is plasma arc cutting apart from shearing as mentioned. In both cases, cut edges will have excessive oxide deposition that is black in color. This is to be removed by grinding, and the acceptable level is at least 0.5 to 1.0 mm deep into the parent metal so that detrimental effects due to the presence of oxides can be completely avoided.

After grind back, cut edges shall be examined visually for any defects such as lamination or any other local defects due to improper cutting parameters resulting from human or machine errors. If needed, visual examination of cut edges may be supplemented by liquid penetrant testing (LPT) and if found satisfactory can be released for subsequent operations such as bending or erection as applicable.

7.7 Bending of Plates

7.7.1 Direction of Bending

Plates have better strength in the direction of rolling (in the mill as indicated in Figure 7.2), compared to its transverse direction. In the case of shell plates, only one plate (closure plate) would be shorter in length; all others would be full-length plates. In the event of full plates, obviously plate bending shall be in the direction of the length of the plate itself. If there is a closure or compensating plate as well, it is preferable that the direction of bending is maintained in the same way, for which longitudinal direction shall be marked on the compensation plate, while marking the component on the full plate.

7.7.2 Prepinging

Plates for the shell may require bending, based on the diameter of the tank and the thickness of the shell courses. API 650 requires no plate bending for shell thicknesses below 5 mm irrespective of the diameter of the tank. Likewise, for shell thickness equal to or above 16 mm (5/8″), bending is required for all diameters. For in-between shell thicknesses, it goes as follows:

- 5 mm (3/16″) to < 10 mm (3/8″) required bending for tank diameters equal to or below 12 m (40′)
- 10 mm (3/8″) to < 13 mm (1/2″) required bending for tank diameters equal to or below 18 m (60′)
- 13 mm (1/2″) to < 16 mm (5/8″) required bending for diameters equal to or above 36 m (120′)

Please refer to Section 6.1.3 of API 650 for finer details.

The bending operation shall start from both ends of each shell plate. Giving the right shape (bending to the required diameter) plays a vital role in maintaining the circularity and true shape of the tank. Therefore, both ends of each shell plate are pressed to the required shape as shown in Figure 7.4. This process is called *prepinging*. After prepinging, the bent profile at both ends is verified using a template, and if satisfactory, full bending is carried out (see Figures 7.4, 7.5, 7.6). Usually, bending is carried out in a few stages to have negligible elongation of shell due to the bending operation. The extent of actual extreme fiber elongation can be measured (if required) by measuring the increase in length of a prefixed

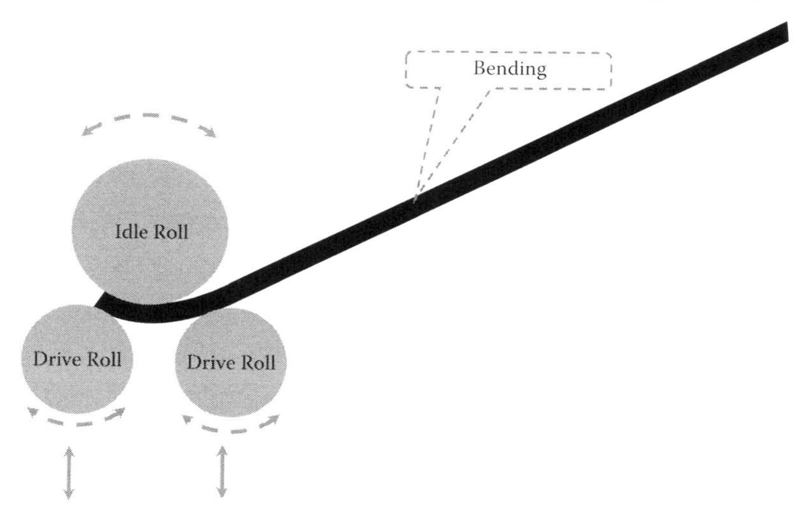

FIGURE 7.4
Schematic of the prepinging operation.

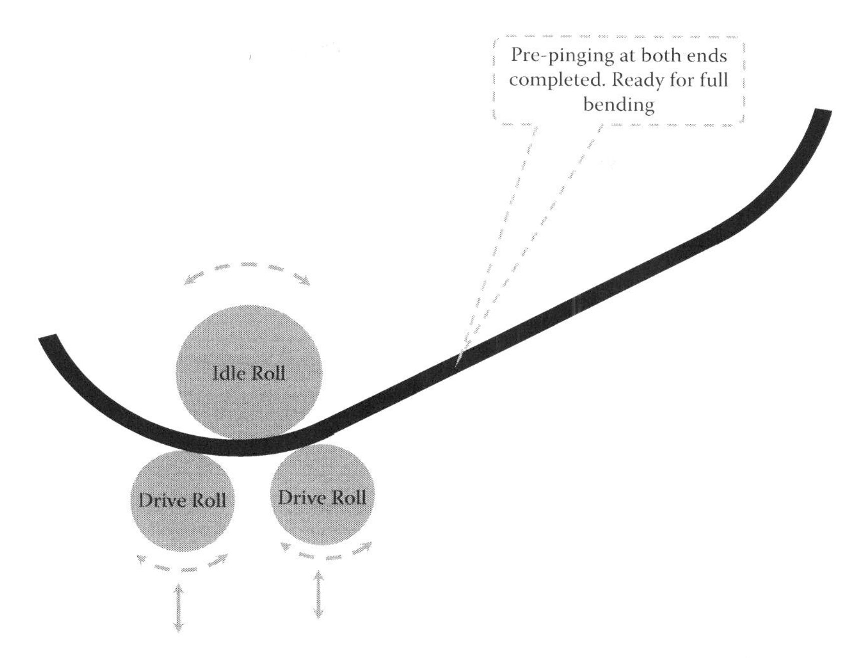

FIGURE 7.5
Schematic of full bending.

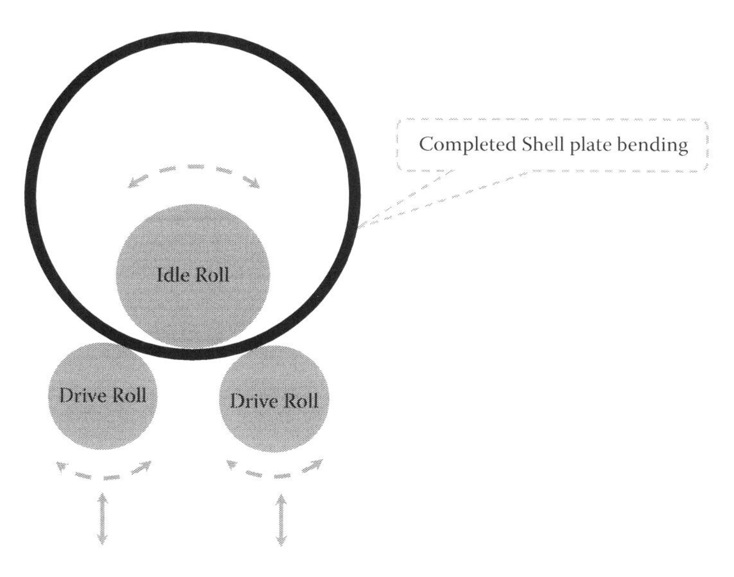

FIGURE 7.6
Schematic of full bending.

distance (say 300 mm) marked on the outside surface of the shell, by providing two punch marks that are to be removed later through dressing. For all normal carbon steels, this would be well below 5%, since formation is taking place only in one direction.

Figure 7.7 shows the progression of the plate bending operation.

The plate is bent toward the top roll by the other roll, and the plate edge is prebent to the minimum flat end possible. A length of the plate is rolled, and the radius is checked.

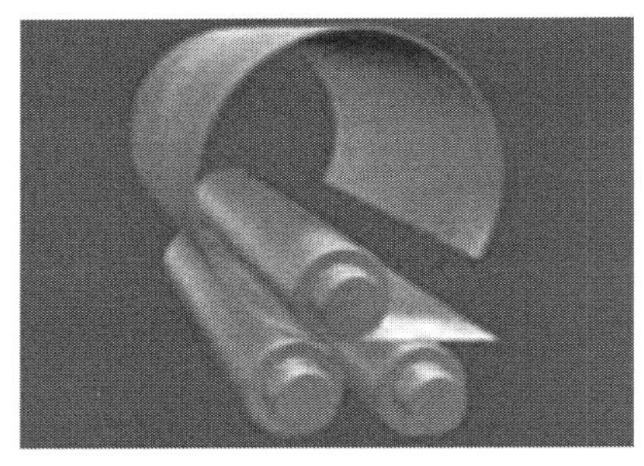

Roll the plate through the machine and bend to the desired diameter.

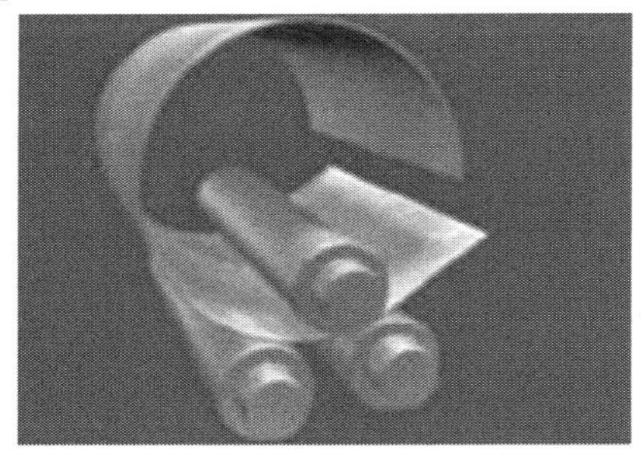

Lower the clamping roll and raise the other roll until the plate is clamped again. Roll the plate into a closed shell and prebend the second plate edge.

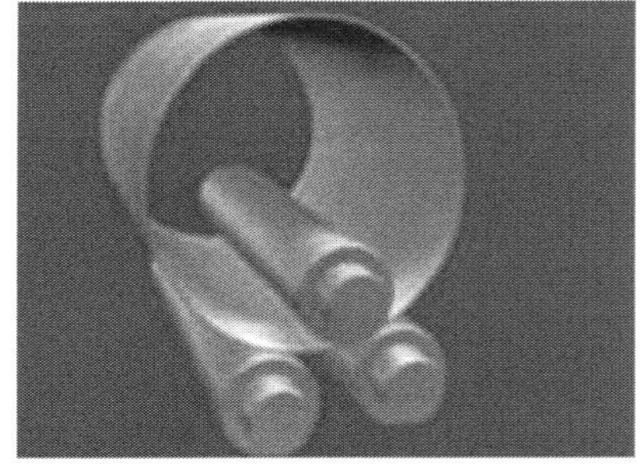

The finished shell is released from the machine by lowering the drop end.

Note: Courtesy of ROUNDO SB, Sweden.

FIGURE 7.7
Plate bending process.

7.7.3 Templates

To check the bending of the shell plate, templates made of aluminum of sufficient thickness and chord length may be used. API 650 is not specific in this regard, and hence it is suggested to follow the template dimensions as per Clauses UG 29.2 and 80 of ASME Section VIII Div (1), with the following limitations.

For large diameter tanks, the Do/t ratio shown in the Y axis will exceed the 800–1,000 range. This automatically calls for a chord length of 0.03 Do as in Clause UG 29.2. The template chord length required as per Clause UG 80 is twice the chord length obtained from UG 29.2 and hence works out to 0.06 Do. For a tank of 100 meter diameter, the chord length of the template works out to 6,000 mm, which makes its handling impractical. Templates with a chord length greater than 2,000–2,500 mm would not be practical, and hence it is recommended to use a chord length of 0.06 Do or 2,500 mm, whichever is the minimum.

The template shall be verified prior to use and shall have the inside diameter or outside diameter (ID or OD) punched on it with the personal stamp of the concerned inspector to avoid its wrong use for other IDs or ODs.

(a)

FIGURE 7.8
(a) Progression of plate bending (pictures courtesy M/s MG s r l, Italy). (*Continued*)

(b)

FIGURE 7.8 (*Continued*)
(b) Progression of thick plate bending (pictures courtesy M/s MG s r l, Italy). (*Continued*)

(c)

FIGURE 7.8 (*Continued*)
(c) Progression of plate bending near completion (pictures courtesy M/s MG s r l, Italy). (*Continued*)

(d)

FIGURE 7.8 (*Continued*)
(d) Progression of plate bending. Methodology to remove bent plate from rolls (pictures courtesy M/s MG s r l, Italy). (*Continued*)

(e)
Plate bending in progress.

FIGURE 7.8 (*Continued*)
(e) Progression of thin plate bending (pictures courtesy M/s MG s r l, Italy).

7.8 Laying of Bottom (Annular, Bottom, and Sketch) Plates

The term *bottom plate* generally includes three types of plates, namely, annular, bottom, and sketch plates, as described next (See Figure 7.9).

7.8.1 Annular Plates

Annular plates are those plates that are placed directly under the shell. As the shell course is placed just on top of this plate, weld joints between two annular

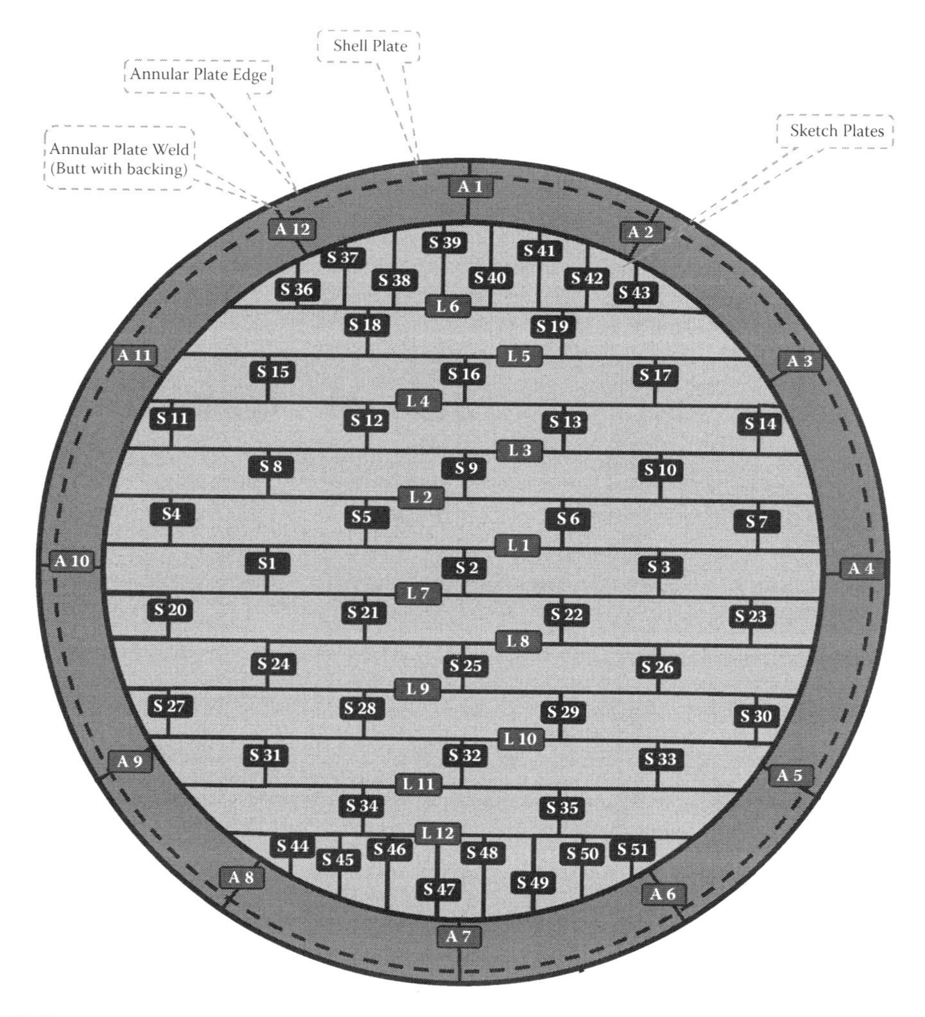

FIGURE 7.9
(See color insert.) Typical layout of annular, sketch, and bottom plates.

plates shall essentially be butt welded with (mostly) a backing strip underneath. Weld reinforcement of this butt weld is ground flush with the parent plate to facilitate proper seating of the shell plate at the location where the shell plate is expected to cross annular plate butt welds. The number and size of annular plates is decided based on the minimum width specified for such plates in client specifications, as well as the available size of plates. A typical arrangement of annular plates, sketch plates, and bottom plates are shown in Figure 7.9.

7.8.2 Bottom Plates

Full-size plates that are used for the tank bottom are designated as bottom plates. Based on the assumed sizes of plates (12,000 mm × 3000 mm), there are 24 such plates, as shown in the bottom plate layout shown in Figure 7.9. They are lap welded to each other with an overlap of about 50 mm to 60 mm welded to the requisite fillet size. These welds are then tested to ascertain its integrity by vacuum box, LPT, or both, in addition to visual inspection, which is mandatory. While welding these lap joints on the bottom plate, utmost care shall be taken to see that distortion is controlled well. A typical welding sequence is provided in Table 7.2 for this purpose with respect to Figure 7.9. In the case of larger diameter tanks wherein more plates come into picture, the staggered welding pattern described can be adapted to suit specific requirements as logical and feasible.

7.8.3 Sketch Plates

Sketch plates are those segmental subsize plates used to bridge the gap between the bottom plate and annular plates. There are 40 such plates present in the configuration of the bottom plate shown in Figure 7.9. These plates are basically polygons composed of a rectangular portion followed by a right-angled triangle, to fill the gap between a regular bottom plate and an annular plate, as shown in Figure 7.9. Depending on the size of the tank and the size of the plates proposed, the configuration of the sketch plate will vary. Sketch plates are also lap welded to the annular plate. As annular plates are welded with a backing strip (between them), invariably the lap of the sketch plate shall be above the annular plate. Section 7.8.9 indicates the welding sequence proposed for the welding sketch plate to the bottom plate, whereas the weld between the sketch plate and the annular plate is taken up only after completion of all main welding of the tank, including the first shell course to the annular plate. The details pertaining to the welding of each type of plate are given separately in ensuing sections.

7.8.4 Annular and Bottom Plate Layout

Once the foundation is completed in all respects including the leveling course with bitumen mix, annular, sketch, and bottom plates are laid sequentially, with required overlap as specified in drawings. (Please note that the underside of bottom plates shall be coated prior to layout to mitigate underside corrosion.) A typical layout scheme for a 54 m diameter storage tank is shown in Figure 7.9. The layout is so designed to optimize the use of plates of size 12,000 mm × 3,000 mm. Annular plates are laid first and butt welded (these are butt welded, usually with a backing strip). As per API 650, these welds require random spot radiography, whereas many client specifications insist for 100% radiography of these joints, especially from the oil and gas industry. While the welding of annular plates is progressing, laying of sketch plate and bottom plates is taken up progressively and tack welded in position.

7.8.5 Surface Preparation and Coating of Bottom Plates

Underside corrosion of bottom (annular, sketch, and bottom) plates is a serious matter, especially when natural phenomena such as a high water table, heavy rainfall, the salt content of soil in the near vicinity, and so on are present. Even otherwise it is better to provide a corrosion-resistant coating prior to laying the plates on the foundation even though cathodic protection is provided for the underside of bottom plates. When plates are coated up to the edge, care shall be taken to use paints or coating materials that shall not produce any detrimental effects in welds affecting the strength of the weld by way of entrapped defects or otherwise. As a precaution, to facilitate the sound welding of bottom plates, usually a 1″ width around the plate edges is left uncoated by applying masking tape before coating.

As in any other coating application, the protective coating also shall be applied after proper surface preparation, as recommended by the paint manufacturer or as specified in the client specification. Being a specialized topic by itself, surface preparation and painting and coating are not dealt with in this book. However, for the sake of completeness, a brief overview of surface preparation and painting and lining is provided in Chapter 13.

7.8.6 Laying of Annular Plates

Before annular plates are laid, the level of foundation prepared by the civil construction team needs to be verified, preferably in a joint examination by both civil and mechanical teams. As per API 650, for the ring wall type of foundation that is very common, the ring wall shall be level within ±3 mm (1/8″) in any 9 m (30′) of the circumference and within ±6 mm (1/4″) in total circumference measured from the average elevation. Similarly, tolerances on the level of foundations for other types of foundations are provided in Clause 7.5.5.2 of API 650.

After getting clearance for annular plate laying, mark the 0°, 90°, 180°, and 270° coordinates and the outside circumference of the annular plate on the foundation from a reference center point obtained through a theodolite survey. Annular plates are to be laid matching the diameter marked on the finished tank foundation, which is to be at the specified datum. Individual plates duly prepared in accordance with the cutting plan mentioned in earlier sections are laid out in sequence to obtain the annular plate around the full perimeter of the tank as shown in Figure 7.9. It is always preferable to maintain the outer diameter of the annular plate at + 5 mm to 10 mm above requirement, to achieve the required diameter after shrinkage of the annular plate welds.

Orientation of annular plate weld joints shall be strictly controlled according to the cutting plan or relevant drawing issued for the purpose. Fit up of annular plate joints shall be carried out using proper jigs and fixtures as per the joint configuration adopted (from code based on feasibility and client requirements, if any) for the tank. Care shall be taken during the fit up, so that there shall not be any gaps between the annular plate and the backing strip (if used).

As the backing-strip joint is one of the most acceptable and extensively used arrangements, a backing strip of requisite size is welded on the rear side of each of the annular plates, and the second plate is kept over the backing strip as per the configuration specified. The sequence of welding of annular, sketch, and bottom plates shall be so designed that distortion resulting from the huge quantity of weld metal deposited on comparatively thin plates shall be minimized so that the resultant undulations or waviness of the bottom plate after welding shall be well within acceptable limits. A typical sequence recommended for the welding of annular, sketch, and bottom plates is provided in Table 7.2 in reference to Figure 7.9. It goes without mentioning that all welding shall be carried out by welders qualified by applicable welding procedure specifications (WPS). It is necessary that all nondestructive testing (NDT) required for annular plate welds need to be completed prior to laying

TABLE 7.2
Welding Sequence

Weld Group	Welds	Sequence	Remarks
	Annular Plate Butt Weld		
A	A 1, A 4, A 7, and A 10	1	Welds in Groups A and B can be welded simultaneously
B	A 3, A 6, A 9, and A 12		
C	A 2, A 5, A 8, and A 11	2	Followed by welds in Group C
	Bottom Plate Short Seams		
A	S 1 and S 3	3	All welds in Groups A, B, C, and D can be attempted simultaneously depending on the number of welders available. Distribute the welding evenly over welds identified.

(Continued)

TABLE 7.2 (*Continued*)

Weld Group	Welds	Sequence	Remarks
B	S 8, S 10, S 24, and S 26		
C	S 15, S 17, S 31, and S 33		
D	S 18 and S 34		
E	S 5, S 7, S 21, and S 23	4	All welds in Groups E, F, and G
F	S 12, S 14, S 28, and S 30		
G	S 19 and S 35		
H	S 4, S 6, S 20, and S 22	5	All welds in Groups H and J
J	S 11, S 13, S 27, and S 29		
K	S 2	6	All welds in Groups K, L, and M
L	S 9 and S 25		
M	S 16 and S 32		
N	S 36, S 43, S 44, and S 51	7	All welds in Groups N and P
P	S 39 and S 47		
Q	S 38, S 41, S 46, and S 49	8	All welds in Group Q
R	S 37, S 42, S 45, and S 50	9	All welds in Group R
S	S 40 and S 48	10	All welds in Group S
	Bottom Plate Long Seam		
A	L 5 and L 11	**11**	The entire length of the weld shall be divided into segments of 1 meter length, and depending on the number of welders available, the segments have to be assigned locations in a staggered manner to the extent possible. Continuous welding of a seam from one end to the other shall be avoided to prevent bottom plate bucking. The alternate stiches also shall be attempted in a staggered manner if possible depending on the number of welders available.
B	L 3 and L 9	**12**	
C	L 4 and L 10	**13**	
D	L 2 and L 7	**14**	
E	L 1	**15**	
F	L 6 and L 12	**16**	
	Sketch to Annular Plate Weld		
A	The sketch to the annular plate joint shall be welded only after completion of the shell to bottom joint welding.	17	The entire circumference is to be divided into segments of 1 meter length. Depending on the number of welders available, the segments have to be allocated at diametrically opposite locations with maximum staggering and progressing in a staggered manner in one direction either clockwise or counterclockwise.

Note: Refer to Figure 7.9.

sketch and bottom plates, since some amount of handling might be required to carry out NDT of annular plate welds, especially for radiographic testing, for which film needs to be placed below.

7.8.7 Recommended Weld Joint Configurations

API 650 recommends weld joint configurations to be followed for all types of joints possible in a storage tank.

In most cases, the weld joint between annular plates shall be a single "V" butt joint with a backing strip. Depending on the thickness of the annular plate, this may be welded together without a bevel as well. Refer to Figure 5.3a (Bottom Plate Joints) of API 650.

Typical bevel edge preparations usually used in construction are shown in Figure 7.10. Beveled joints are preferred to unbeveled joints, even for lower thicknesses. However, for weld joints with a backing strip, bevel joint design 7.10(a) or (c) shall be possible, since welding from one side alone is possible in this instance. For higher thicknesses, the bevel design in Figure 7.10(b) is the ideal from a distortion point of view, especially when unequal V of 2/3 and 1/3 are provided from the face and root sides, respectively, and back chipped and welded from the root or second side.

Whatever the type of weld adopted, annular plate joints welding shall be done by welding alternative joints in a staggered manner detailed in the welding sequence in Table 7.2 together with Figure 7.9. Qualified welders to applicable WPS shall be deployed for welding. In case any defects are found during welding, the weld shall be removed by grinding, followed by a dye penetrant check to ensure the removal of the defect. The joint shall be welded back to the required size and inspected again as required in the code or specification. Completed welds shall be cleaned using a manual or rotary wire brush or by grinding. The jigs and fixtures used for fit up annular joints shall be removed on completion of welds, and tacks or attachment welds made for this purpose (if any) shall be removed by grinding. Locations where temporary welds were made shall undergo satisfactory MPT.

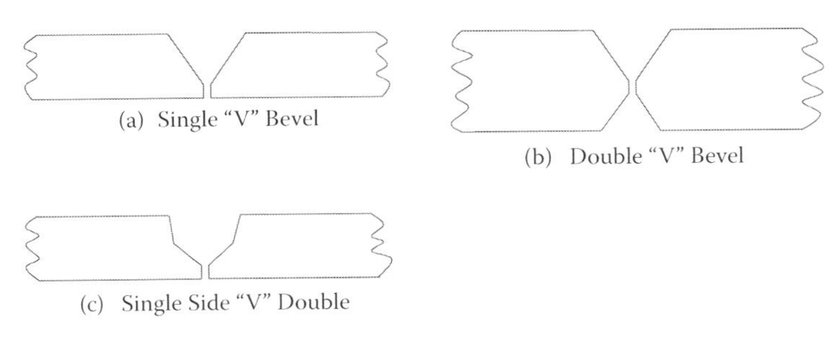

FIGURE 7.10
Typical bevel edge preparations (for butt joints).

API 650 recommends a few types of weld joint configurations based on the thickness of the bottom plate and shell course thickness. They are provided in Figure 5.3a (Bottom to Shell Joint) and in Figure 5.3c (Detail of Double Fillet-Groove Weld for Annular Bottom Plates with a Nominal Thickness Greater Than 13 mm (1/2″)) of API 650. The minimum sizes of fillet welds applicable to Figure 5.3c is provided in the table given in Clause 5.1.5.7 of API 650.

7.8.8 Laying of Bottom and Sketch Plates

The next difficult task is to lay sketch plates. Usually the sketch plates lap is provided below the annular plate, as shown in Figure 7.11(a). Since annular plate welds are carried out with a backing strip, aligning sketch plates is little tricky. To have a proper weld of the sketch plate to the annular plate, the backing strip provided below the annular plate is not provided up to the

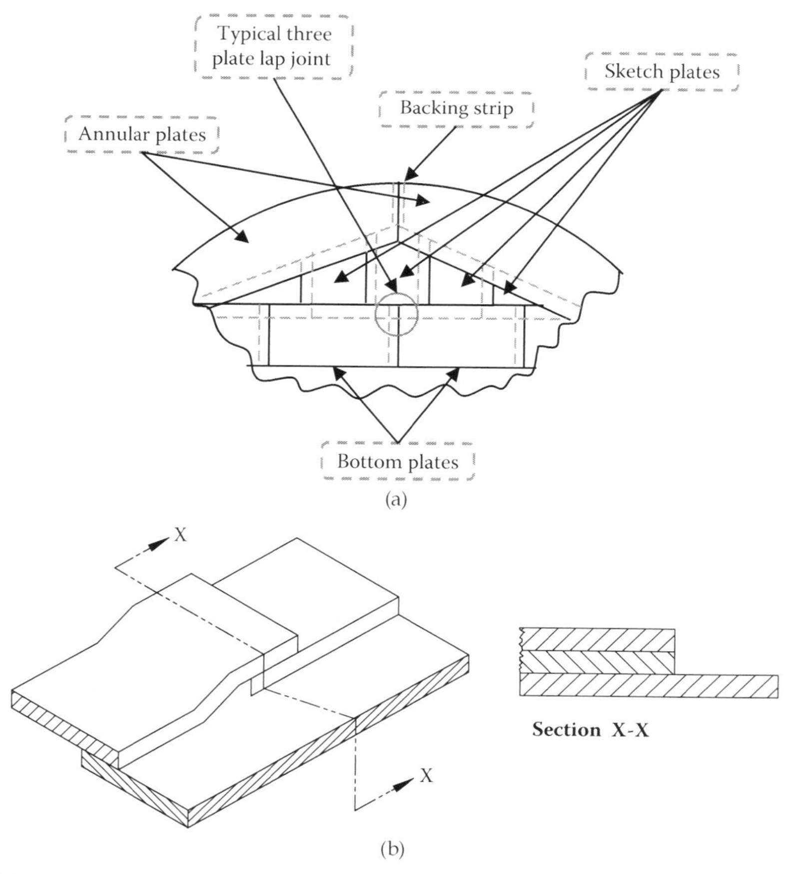

FIGURE 7.11
Plan view of (a) bottom plate lap joints. (b) Three plate lap joint.

inner edge of the weld. The backing strip usually shall be 50 mm shorter (the usual lap provided in lap welding) than the weld length of the annular plate, which is usually left unwelded till the sketch plate welding, which is taken up only at the last stage.

Upon completion of laying all sketch plates as per design, bottom plates are laid according to the required lap indicated in the drawings. The bottom plate coming at the center needs to be laid first and shall progress in all directions. Laps as required (50 mm to 60 mm usually) shall be maintained while fitting up short and long seams as per the drawing. Temporary tack welds may be required on short and long seams to avoid unwanted movement, so also to avoid excessive gaps between the plates at laps.

Short and long seam welds shall be carried out in sequence as indicated in Table 7.2 to minimize distortion of the bottom plate during welding. Joggling shall be carried out by heating and hammering wherever three plate lap joints (see Figure 7.11(b)) are to be made.

Sketches shown in Figure 7.12 show a typical bottom plate configuration under shell in two different scenarios.

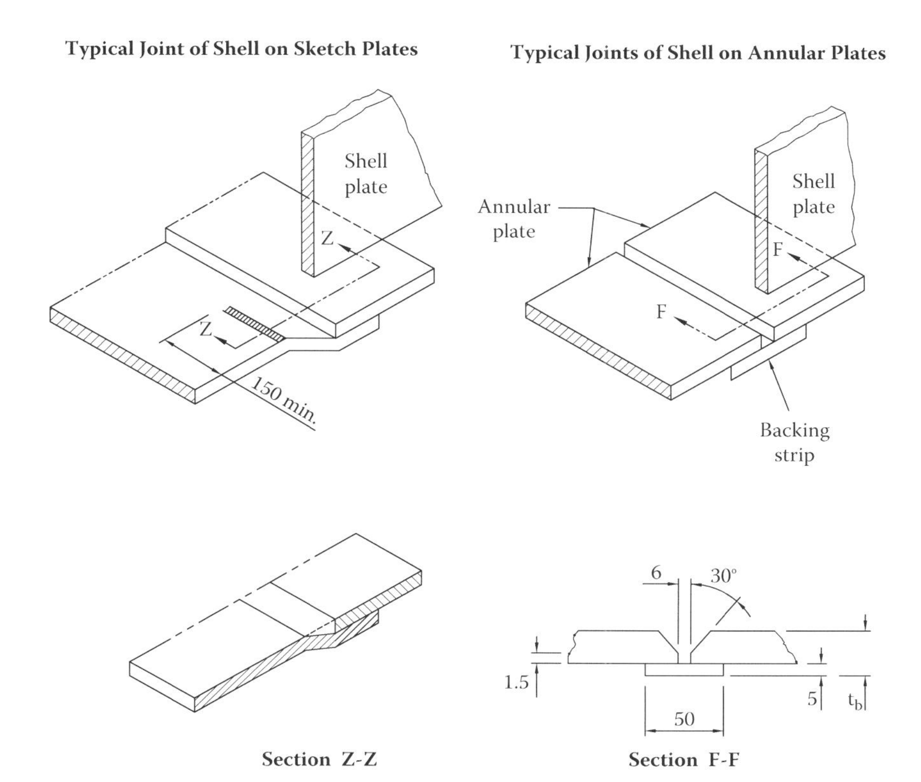

FIGURE 7.12
Typical lap preparations under shell.

7.8.9 Welding Sequence of Annular, Sketch, and Bottom Plates

With reference to the bottom plate layout shown in Figure 7.9, the welding sequence for a different category of welds is developed with a view to reduce distortion of the bottom plate. For easy understanding, the welding sequence is presented as Table 7.2 with sequence and subsequences.

Welds identified in Figure 7.9 are segregated into groups to be taken up in sequence and are indicated in the blue column. The short seam width is considered to be 3 meters, based on the maximum width of the plate usually available in the market and also the maximum plate bending width commercially available. The entire length of welds (3,000 mm) in short seams shall not be welded in a single stretch. Instead, the weld may be divided into three segments of 1,000 mm each and shall be welded at the middle (1 m length marked at the middle) first and then on either end.

The very same methodology shall be adopted for long seams as well, wherein the maximum length expected would be to the tune of around 47 meters for the tank size considered here. This methodology shall be applicable to annular to bottom plate welds as well and would be preferable for any weld of length of more than 1 meter to 1.5 meters in length.

In the case of annular to sketch plate welds (to be taken up as the last weld in the bottom plate), a lot of care is required. In the case of a tank of about 50 meters in diameter, the approximate weld length shall be around 150 meters. When the weld is divided into six segments over the circumference, each segment shall contain a length of 25 meters of weld. If a welder is allocated for each segment, he or she shall weld the same in a staggered manner. The segment allocated shall be subdivided into portions that are 1 meter long. Starting from the middle portion, the welder shall progress in both directions (within the segment), welding alternate portions on either side of the designated midpoint. Later, the welder has to come back and complete the remaining segments of welds allotted to him or her in a progressive but staggered manner.

7.8.10 Annular Plate Welds

Annular plate butt welding shall be carried out in accordance with an approved welding procedure. The process for this welding is selected based on various constraints such as the time schedule, availability of welders, skill level of welders, quantity of welding required (welding economics), and so on. Since the usual width of annular plate butt welding is in the range of 2 meters to 2.5 meters, many times the shielded metal arc welding (SMAW) process is selected for these welds. Depending on the annular plate thickness required, the joint configuration may or may not be provided with a "V," as mentioned in Section 7.8.7. A square butt weld joint configuration is generally acceptable up to 6 mm thick annular plates. Beyond this thickness, either a single or a double bevel is recommended. A double bevel is usually used when the plate thickness is above 16 mm.

In the case of annular plate welds for storage tanks, a double bevel is usually not possible on account of the lack of precision handling equipment at the site. Therefore, even if the thickness of the annular plate exceeds 16 mm, a single bevel is resorted to on account of practical limitations.

As mentioned earlier, codes and good engineering practices do not recommend any beveling for plates below 6 mm thickness. However, it is always better to have a bevel for plate joints, which will facilitate easy flotation of the slag to the top, thereby reducing the chances of slag inclusions and lack of side wall fusion in welds.

Normal tests that are carried out after welding the annular plates are visual examination, spot radiography, and penetrant testing at the areas where the reinforcement is ground flush with the parent plate to facilitate proper seating of the shell plates. Even though these are not mandatory requirements, in the best interest of good quality work, and to avoid reworks at a later stage, it is better to carry out the penetrant test of all butt welds of the annular plate from the outside. Visual examination is the most powerful tool among inspections, and it can rule out a lot of discrepancies in welds and hence in no case shall be compromised. It should be carried out meticulously through every inch of weld on a storage tank.

7.8.11 Bottom Plate Welds

For joints between bottom plates, lap welding is proposed as shown in Section 7.8.8. In this instance tack welding at the underside is not possible and hence needs to be done on the surface to be welded. Even for this lap welding, it is better to remove tacks entirely during welding of the joint, though code does not prohibit its adoption into the main weld, provided it is carried out by a qualified welder using a qualified WPS and compatible electrodes.

The welds between bottom plates are simple lap welds. The completed welds shall be inspected visually and then given a vacuum box test. In the vacuum box test, the amount of vacuum applied and amount of overlap given during testing plays a very vital role in ensuring the quality of the bottom plate weld and hence shall be carried out meticulously. The code specifies the methodology to carry out vacuum box testing, and the details are provided in Chapter 10.

7.8.12 Weld between Annular and Bottom Plates

The major portion of this weld is a simple lap weld with an overlap of about 75 mm instead of the usual 50 mm for other bottom plate lap welds.

However, at locations where long seams of bottom plates are coming, a three plate lap shall eventually occur, and hence care shall be taken to maintain proper alignment of plates while welding so that the gap between plates is minimal. Furthermore, the weld shall cover the entire three lap portion and bent portion of the bottom plate. These points are the most likely areas for a leak in vacuum box testing and hence need extra care during welding and to be inspected thoroughly afterward.

Tests applicable to bottom plate welds are applicable here also. However, in the event of three plate lap joints, where joggling is required to match double lap, welds around the double lap shall be compulsorily dye penetrant tested. This could reveal not only the flaw in the weld but also any flaws that might have occurred during the cold-hot joggling operation carried out.

7.8.13 Testing of Welds

Upon completion of welds pertaining to each category, they shall be subjected to tests as specified in the code and also according to client specifications. The code always specifies the minimum amount of testing that usually would be increased by clients based on their past experience, which shall be reflected in the technical procurement specification (TPS) for the tank. As it is not practical to give all this information in bits and pieces to the construction group, all these technical conditions (other than those specified by the code) are essentially to be listed out in the general arrangement drawing, so that the construction group need follow only the drawings and code, which it should be familiar with.

Apart from normal NDT methods, other effective inspection methods employed in storage tank erection are visual examination, vacuum box testing, pads air testing, and the final hydrostatic/pneumatic testing (of the roof).

Butt weld (usually with a backing strip) is proposed for annular plate welds to facilitate easy erection and for sound welding of the annular plate to the bottom shell course. As per code, the NDT required for this weld is only random spot radiography, apart from visual examination, whereas many client specifications call for 100% radiography. Since full penetration is required for this weld, it is logical to ask for radiography; however, the chances of unacceptable defects in butt weld of a few passes, that too with a backing strip is highly unlikely. Therefore, to ward off untoward incidents such as leaks in the annular plate to shell welds during a hydrostatic test, it is better to carry out the vacuum box test of the annular plate butt weld, in addition to spot radiography (usually specified by clients) and LPT as complementary tests. This test is not very expensive but can effectively locate any pinhole kind of defects that might have been missed in other inspections.

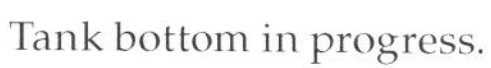
Tank bottom in progress.

8

Erection of Shell Courses, Roofs, and Other Appurtenances

8.1 General

For the erection of storage tanks, two well-established methodologies are available. The earliest methodology was the conventional erection of a storage tank from the bottom to top shell course (one after another) and roof structure and roof after that. The jacking methodology was developed later, wherein the erection of the shell courses start in reverse order from the top to bottom shell course. Each has its own advantages and disadvantages. The jacking up method has advantages in the case of fixed cone roof tanks, whereas for floating roof tanks, structural rigidity at the top could be a concern during lifting in some cases. Furthermore, there is no advantage of welding undercover for floating roof tanks, which depends on how and when the floating deck is constructed.

As mentioned earlier, in the conventional method, shell courses are erected one after another, starting from the first shell course to the top-most shell. Therefore, welding at height is required as the tank acquires height. Moreover, wind would be more at height, and hence the quality of the weld could be a concern for welds carried out at heights. In the jacking up method, the top shell course is erected first, followed by the roof structure and roof plates. Later, other shell courses from top to bottom are erected one after the other. The main advantage is that the majority of welding can be performed at a height slightly (500 mm to 1,000 mm) above the shell course width, which is of much significance from safety and convenience points of view.

8.2 Conventional Method

In the conventional method of the erection of storage tanks, shell courses are erected from the first shell course to the top shell course from the bottom of the tank in a progressive manner. Upon completion of laying the bottom plates (inclusive of annular plates, bottom plates, and sketch plates), the outer and inner circumferences of the tank are marked on the annular plate. Wherever this marking crosses butt welds (with a backing strip) of the annular plate, weld reinforcement is ground flush with the parent metal to facilitate proper seating of the shell to the annular plate. This flush grinding may be done for a width of about 25 mm (minimum) on either side of the shell thickness already marked on the annular plate. Furthermore, it is presumed that all welding between the bottom and sketch plates is completed prior to the erection of shell courses.

Shell plates shall be erected as per the cutting plan and be carried out based on the orientation marking on the annular or bottom plate. In fact, laying the bottom plate shall be done according to this orientation so that once the bottom plate is laid, the orientation cannot be changed. The shell plates of the first course shall be erected one by one progressively till the entire circumference (except the closure plate) is completed. Subject to satisfactory inspection of fit up, vertical seams can be welded. After welding and completion of satisfactory visual inspection of these welds, they are released for subsequent nondestructive testing (NDT) as required by code and specifications.

Upon completion of the weld and NDT of vertical seams, the second shell course erection can be started as in the case of the first shell course, with the only difference being that erection is taking place at a height of one shell course above the bottom plate and foundation. Hence, a proper scaffolding arrangement shall be provided for easy and safe movement of personnel. The scaffolding arrangement can be of two types, one that is purely temporary, erected for the construction purpose without providing any welded cleats on the shell, and the other with scaffolding cleats permanently welded to the tank surface (both inside and outside) if permitted by the client. The purpose of permanent erection cleats is to reduce scaffolding cost and scaffolding erection time during maintenance inspections, to follow during the service life of the tank. In such cases, these cleats are formed as specified in the design code and specifications and data sheets and shall be welded to the shell at proper predesigned intervals and elevations by a qualified welder using approved welding procedure specifications (WPS). This welding also shall be subjected to visual inspection and NDT as required in code and specifications and shall be free of open defects such as undercuts, blow holes, and so on that could act as an eventual stress riser, thereby causing corrosion during service.

During the welding of the vertical and horizontal seams of a storage tank, distortion associated with welding is one of the issues to be tackled based on practical judgment. Weld distortion is an inherent evil associated with every welding that cannot be eliminated under any circumstances. However, if proper techniques are adopted, distortion can be controlled to a great extent. For bottom plate welding, the sequence developed in Table 7.1 is with this intention based on practical experience at the site and hence recommended for all types of tank constructions (conventional and jacking up) in general. As mentioned therein, the weld between the annular plate and the bottom plate (consisting of sketch and bottom plates) is welded only after completion of the annular plate to shell plate welding. Similarly, for shell welds (both vertical and horizontal seams), the welding sequence proposed is explained in Section 8.5.2 and in Table 8.2.

8.3 Bygging or Jacking Up Method

In the jacking up method, the methodology adopted to lay the bottom plate is the same, whereas the erection of the shell plates takes place in reverse order, which means shells are erected progressively from the top to bottom course. The methodology followed by most contractors is explained briefly in the following paragraphs.

On completion of laying the annular, bottom, and sketch plates, as done in the conventional method, the inside and outside circumferences of the first (bottom-most) shell course shall be marked on the annular plate. Thereafter, stools of 400 mm to 500 mm height with a rectangular surface plate at the top of the stool shall be placed along the marked circumference. The stools shall be preferably made of pipe with a size in the range of 350 to 450 (12″ to 16″) NB. As a rule of thumb, stools shall be placed at a distance of about 1.5 m along the circumference. This arrangement shall be capable of taking up the full load of one shell course due for jacking. Afterward, the top surface of the square plate provided on the stools shall be leveled using the water level, and the circumference of the top-most shell course shall be marked on it. Subsequently, top shell plates, bent to a right curvature, shall be placed on the stools as per the shell cutting plan issued for this purpose, with proper orientation as per the drawing. Vertical seams are then aligned, inspected, welded, and visually inspected again. NDT as envisaged in the code and specifications and in the drawings shall follow. Once this is clear (or while this process is going on, as approved by the client), the top shell is now ready for fitting up of the roof structures.

Before the roof plates are placed, the structural support for the roof shall be ready. Therefore, the first step shall be to assemble and weld the curb angle or compression ring (as the case may be) and also to erect the structural members intended to support the roof plate and other appurtenances that are to be installed on the roof. Structural members are usually supported at chairs provided on the shell that in turn are strengthened by the compression ring or curb angle or by reinforcing pads and finally connecting through bolting. Once all necessary bolting of the structure as per the design is completed, structural members shall have the required strength to carry roof plates and appurtenances. Thereafter, the roof plates are laid in the same way as the bottom plates. For the welding of these plates, proper sequence shall be adopted so that adverse effects due to distortion can be kept at a minimum possible level so that water logging on the roof due to buckling of roof plates can be avoided. Because of this issue, roof plates shall never be welded in excess of what is being stipulated in the drawing, which may lead to excessive distortion. Welding of the roof shall be followed by visual inspection, as well as other NDT specified in the code, specifications, and drawings.

On completion of the roof, roof appurtenances shall be welded to the roof as per specifications and drawing. Manufacture, fit up, and welding of these items are dealt with separately under subassemblies in the following sections. Along with this work, the roof and top shell course, platforms, portions of stairways, ladders, and other supporting structures for piping and so on shall also be completed prior to lifting the first shell course, which in reality is the top shell course.

The next step shall be lifting the top shell course up to the height of one shell course to facilitate placement of the second shell course from the top. To reduce the erection time, shell plates for the second shell course from the top shall be placed in position along the circumference to start fit up immediately upon lifting the top shell course. Initial lifting shall be to a height of +25 mm above the next shell course height so that the shell course plates are erected on a packing plate placed above the stool. Later, on completing the placement of all the plates and its tack welding, the same shall be lowered to the stool by removing the packing plates provided, which makes the lifting and erection quite easy and smooth. This process is repeated for all subsequent shells till the bottom shell course is reached, by which time the lifting arrangements can be removed. Though the above mentioned methodology is the ideal, quite often first lifting of the shell was carried out after erection of two shell courses at the top. This is to provide ample head room for personnel working inside the tank even after the erection of roof structural supports which could be quite big depending on the diameter of the tank. In such cases, the first erected shell course shall be the top most but one shell course, followed by the top most shell in conventional manner. Thereafter the roof supports, roof plates and other attachments on roof and these two shell courses are completed. Therefore the first jacking will be the top two shell courses and roof together. All subsequent lifts shall be after addition of one shell course each.

By the time erection reaches the bottom-most shell, stools are removed, and any damages or dents that occurred on annular or bottom plates during the erection of shells shall be corrected satisfactorily. The shell plates shall then be placed on the annular plate, and its alignment shall be carried out. The vertical joints are welded first, and later the shell is welded to the bottom plate joint, both inside and outside.

8.4 Comparison of Erection Methodology

In the conventional method, as work is progressing from first to top shell courses, a large amount of work including welding shall be taking place at heights. This calls for scaffolding and stringent safety requirements for all working personnel. Since a large quantity of welding is taking place at height, the probability of weld defects is higher on account of lower supervision, lesser accessibility, adverse wind conditions, and so on. In addition to possibly affecting the running plants in the near vicinity, welding at heights poses hazardous threats to existing storages, especially in the case of tank farm extension projects. In such instances, the obvious choice in methodology shall be the jacking up method.

On the contrary, some of the negatives that can be attributed to the jacking up method include extra cost, precautions needed during jacking up and erection, probable damages that may happen to the shell because of misaligned jacks, and oil spillage on the bottom plate from leaking jacks. In spite of these issues, this method is gaining momentum, especially in the context of the high degree of concern given to the safety of working people and the high quality standards required for storage tanks.

	Advantages		
	Conventional Method for Erection		**Jacking Up Method for Erection**
1	Erection work does not require heavy equipment or such facilities.	1	Welding operations are executed either at ground level or at lower heights, thereby providing a conducive atmosphere for better quality in welding and workmanship in general.
2	Services of highly skilled operators are not necessary, as erection is carried out plate by plate.	2	Work at lower heights reduces risk during construction.
3	The erection operation is cheaper compared to that of the jacking up method.	3	There is easier access to and better control over welded joints.
		4	Scaffolding is not required.
		5	Since the tank roof is constructed first with all its paraphernalia, complementary stabilizing of the shell against loss of stability in radial direction caused by wind pressure during erection is not necessary.
		6	The major part of welding can be carried out under the erected roof.
		7	The protective tank roof eliminates wind damage during welding of the shell.
	Disadvantages		
	Conventional Method for Erection		**Jacking Up Method for Erection**
1	A relatively longer term for erection is needed.	1	It is costly.
2	The number of weld joints and necessary control on site is increased.	2	The hydraulic jacking system for the entire fabricated weight of the storage tank except the first shell course and the bottom plate needs to be mobilized.
3	More qualified workers are needed for this operation.	3	Mobilization and set up time is more.
		4	Work on each stage shall be completed (including coating); if not, scaffolding for the application of coating shall be needed again.

8.5 Erection of Shells by Conventional Method

The conventional method of storage tank shell erection consists of following distinct steps.

8.5.1 Shell Cutting Plan (Shell Development)

A typical shell cutting plan for a storage tank is shown in Figure 8.1. The shell cutting plan shall be the same irrespective of erection methodology. Shell development is made using available sizes of plates in the market and also considering the maximum width of plate that can be accommodated in the plate bending machine. Based on the layout pattern of the bottom plate, orientation of shell vertical joints also shall be indicated in the shell cutting plan with positions of nozzles and other attachments. This will help in identifying probable fouling of attachments with weld seams, as well as fouling with each other. If the position of other attachments is still not frozen, then at least the orientation and elevation of nozzles shall be marked to verify their fouling with weld seams, especially those with the vertical seams of the tank.

8.5.2 Erection of Shell Courses

To erect the shell on the annular plate, it is preferable to complete the following works, including testing welds as specified in the codes, specifications, and drawings:

- Laying of annular plates, welding between them, its inspection and NDT

Diameter	Height	Plate size	Type of joining of shells
38,000 mm	20,000 mm	12,000 mm × 3000 mm	Inside flush

Shell thickness	Shell 1	Shell 2	Shell 3	Shell 4	Shell 5	Shell 6	Shell 7	Shell 8
	22 mm	20 mm	18 mm	12 mm	10 mm	8 mm	8 mm	8 mm

H8	S8 V1	S8 V2	S8 V3	S8 V4	S8 V5	S8 V6	S8 V7	S8 V8	S8 V9	S8 V10	
	S7 V1	S7 V2	S7 V3	S7 V4	S7 V5	S7 V6	S7 V7	S7 V8	S7 V9	S7 V10	H7
H6	S6 V1	S6 V2	S6 V3	S6 V4	S6 V5	S6 V6	S6 V7	S6 V8	S6 V9	S6 V10	
	S5 V1	S5 V2	S5 V3	S5 V4	S5 V5	S5 V6	S5 V7	S5 V8	S5 V9	S5 V10	H5
H4	S4 V1	S4 V2	S4 V3	S4 V4	S4 V5	S4 V6	S4 V7	S4 V8	S4 V9	S4 V10	
	S3 V1	S3 V2	S3 V3	S3 V4	S3 V5	S3 V6	S3 V7	S3 V8	S3 V9	S3 V10	H3
H2	S2 V1	S2 V2	S2 V3	S2 V4	S2 V5	S2 V6	S2 V7	S2 V8	S2 V9	S2 V10	
	S1 V1	S1 V2	S1 V3	S1 V4	S1 V5	S1 V6	S1 V7	S1 V8	S1 V9	S1 V10	H1

FIGURE 8.1
(See color insert.) Shell cutting plan with weld map.

- Laying of bottom plates, welding between them, its inspection and NDT
- Laying of sketch plates, its welding with bottom plate, its inspection and NDT

To minimize buckling of the bottom plate, the weld between sketch plates and annular plates is done only at the end, that is, at any time after welding (inside and outside) of the first shell course to the annular plate.

On the completed annular plate, the outer and inner circumferences of first shell course of tank are marked. Wherever this marking crosses butt welds (with a backing strip) of the annular plate, weld reinforcement is ground flush with the parent metal to facilitate proper seating of the shell to the annular plate. This may be done for a width of about 25 mm on either side of the shell thickness already marked on the annular plate. For storage tanks, shell thickness shall vary with height, meaning the first shell course would be thicker compared to others due to variation in static head envisaged.

Each shell course shall be erected as per the cutting plan in relation to the orientation marking already made on the bottom plate.

Plates of the first shell course shall be erected one by one progressively till the entire circumference is completed. After fit up of entire shell plates, the circumference at both ends (at top and bottom) is recorded. This reading shall be recorded under the head "During Fit Up" with weld gaps as provided. In practice it is observed that whatever gap provided at fit up vanishes after welding because of shrinkage of welds. Therefore, it is recommended that the shell circumference shall be maintained at the actual circumference required + gap X number of vertical seams, worked out as follows:

Required outside circumference according to drawing	= (ID + 2 T) X ∏ mm say X
Number of shell plate required for the shell course	= 10
Therefore, the number of vertical seams	= 10
Gap specified in weld joint configuration	= 3 mm
Therefore, outside circumference at fit up	= X + 10×3 mm = X + 30 mm

For the typical tank shell development shown in Figure 8.1, the circumference requirement during fit up with a gap of 3 mm (as in design) is worked out as shown in Table 8.1.

The circumference shall be measured at both ends (top and bottom) of each shell course at about 50 mm to 75 mm away from the edges (horizontal joint). If these measurements are within acceptable limits, vertical joint edge preparation shall be checked for compliance to the drawing and codes. Further profile at the weld joint also shall be checked, and correction if needed may be done prior to the start of welding. When all these parameters are within allowable limits, vertical seams can be released for welding.

TABLE 8.1

Shell Circumference

Shell Course	Shell 1	Shell 2	Shell 3	Shell 4	Shell 5	Shell 6	Shell 7	Shell 8
Shell thickness (mm)	22	20	18	12	10	8	8	8
O/S circumference (mm)	119,519	119,506	119,494	119,456	119,443	119,431	119,431	119,431
O/S circumference with gap (mm)	119,549	119,536	118,524	119,486	119,473	119,461	119,461	119,461

Welders qualified to applicable WPS shall carry out all welds between pressure parts and welding that connect a load-bearing member to the tank. In the event of thick shells (above 16 mm), where double "V" edge preparation is required, welding may be carried out simultaneously from both sides (with a little time difference to have sufficient offset of molten weld puddle), inside and outside. The recommended sequence for welding of shell seams is shown in Table 8.2.

The sequence can be adopted for vertical joints in all shell courses. Many times, more welders may be required to meet time schedules. In such cases, welds identified in each lot shall be attempted in a distributed or staggered manner. For instance, if two welders are deployed, they can attempt V 2 and V 6 initially followed by V 4 and V 8, to maintain staggering of seams welded at a point in time.

On satisfactory completion of welding, seams are subjected to visual examination for undercuts, underfill, uniformity, reinforcement, weaving, surface pores, and profile at weld (peaking in or out). If seams are found acceptable on all these aspects, they are released for NDT, as specified in the code, specifications, or drawings, whichever is most stringent.

Upon satisfactory completion of NDT, the fillet weld between the annular plate and the shell is to be carried out. It is recommended to adopt a sequential or staggered welding for this joint, as this could have a considerable impact on the resultant buckling of the bottom plate. The sequence of welding proposed for sketch plate to annular plate welding provided in Table 7.2 may be followed here also. Welding on both sides (inside and outside) may be taken up simultaneously with a little time difference to avoid concentration of heat at one location.

Extreme care shall be taken while welding the shell to the annular plate joint to provide the right size of weld so that detrimental defects that may get entrapped in excess weld and related distortion due to excess weld can be avoided. At the same time, this weld being the most stressed joints of a

TABLE 8.2

Shell Vertical Seam Welding Sequence

First lot	V 2	V 4	V 6	V 8	Weld from bottom to top	For double "V" joints, it is better to weld simultaneously from inside and outside so as to reduce distortion.
Second lot	V 3	V 5	V 7	V 9		
Third lot	V 1	V 10	[a]			

[a] For access into and out of the tank, and moving large components as well, the compensating segment of the first shell course is fitted and welded as a last operation.

storage tank, every location in this weld shall have the required minimum size of fillet as well.

On completion of the weld, visual inspection shall be carried out under sufficient light, especially during inspection of the inside weld. Depending on the service of the tank, this visual inspection may be supplemented by penetrant testing, magnetic particle testing, chalk oil test, and so on as per code and client specifications.

When all above-mentioned works are completed satisfactorily, the tank is now ready for the erection of the second shell course. Erection of the second shell is carried out in the same way as that of the first shell, with surfaces of plates either just matching or with a fixed offset in case of the difference in shell thicknesses. The only difference is that here horizontal seam No. H1 and the vertical seams of the second shell course are welded simultaneously. Furthermore, temporary scaffolding is needed at erection height for easy alignment of the second shell course. As the shell has to be checked from both inside and outside, scaffolding shall be provided on both sides. Temporary cleats (if any required) welded to the shell for erection of any temporary platforms shall be removed by grinding and checked by magnetic particle testing or dye penetrant testing as required in the standards, specifications, or drawings.

This methodology is repeatedly adopted for all subsequent shells courses till the tank reaches its full height.

Whenever the tank reaches shell-carrying wind girders and other structural supporting members, and after welding of the vertical seams and related NDT, these structural members are erected to provide structural stability to the tank. Till the erection of wind girders, the tank shall be provided with temporary ties using guy wires to circumvent probable buckling collapse or toppling that may occur because of heavy winds. Therefore, it is important to provide these supports at appropriate heights at the earliest opportunity as proposed by the design.

Subsequent shell sections are also erected in a progressive manner, level after level. Some shell sections may also have to be provided with intermediate wind girders. If so, they are also to be erected prior to erection of the next shell course. Parallel action to prefabricate wind girders is absolutely essential to achieve the desired progress. Shell erection in the above-described manner shall be continued till the tank reaches the top-most shell.

For welders to have proper control over welding the horizontal seams, the following sequential approach is recommended. For working out the sequence, as in the case of vertical welding, it was presumed that five welders would be deployed for each horizontal seam as well. Since the total circumference in this instance works out to 120 m, welders are assigned zones of 24 m along the circumference, irrespective of whether they are welded from the inside or outside. In this instance, it is preferable that the same welder completes the entire thickness of weld in the assigned zone, which is subdivided into segments as mentioned in the following table and which would be helpful in controlling the quality of welds produced by each welder.

Horizontal Seam Welding Sequence

Every Zone of 24 Meter Weld of Total 120 Meters Welded by One Welder					
Seam	**First Segment**	**Second Segment**	**Third Segment**	**Fourth Segment**	**Fifth Segment**
H2 (S1/S2)	W1	W2	W3	W4	W5
H3 (S2/S3)	W2	W3	W4	W5	W1
H4 (S3/S4)	W3	W4	W5	W1	W2
H5 (S4/S5)	W4	W5	W1	W2	W3
H6 (S5/S6)	W5	W1	W2	W3	W4
H7 (S6/S7)	W1	W2	W3	W4	W5
H8 (S7/S8)	W2	W3	W4	W5	W1
H9 (S8/Curb <)	W3	W4	W5	W1	W2

Note:
1. Every zone thus assigned is further divided into three equal segments of 8 m weld each.
2. The first welding of each welder shall be the middle segment (8 m segment). This will provide a sufficient working gap between welders.
3. After they complete the first segment, all welders shall attempt the segment on their right.
4. As the last, the segment on the left shall be taken up.
5. Complete welding from inside and outside, done in a balanced manner for each segment.
6. Welders move to segments on the right and left only after completing welding of each segment.
7. Zone and segments shall be identified properly against each welder deployed.
8. All welding within designated segments shall progress in either clockwise or counterclockwise direction.

Erection of first shell course in progress.

Erection of first shell course in progress (nearly complete).

Erection of second shell course in progress.

Erection of shell at higher elevation in progress (view from inside).

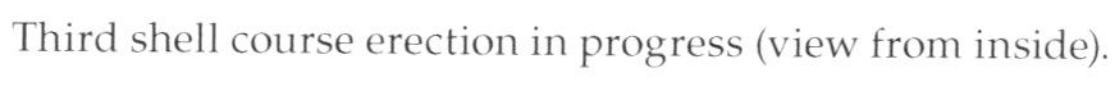

Third shell course erection in progress (view from inside).

Fourth shell course erection in progress (view from outside).

8.5.3 Erection of Roof Structure and Roof Plates

For the top-most shell, on completion of welding of vertical seams, compression ring and structural support seats are fitted as per design. They are to be welded first to impart the required strength to the top shell course of the tank, followed by the erection of roof structures. This is carried out with the help of temporary structures (consisting of a fabricated column) right up to the place where the central support drum is expected to be located in elevation and plan. In the event of self-supporting structures, upon completion of erection of the entire roof structure, the temporary structure provided to erect the central support drum is removed.

Roof plates are laid on the roof structure, as required in the drawing. Here also, the welding sequence and weld quantity play a vital role in determining the final contour of the roof plate. Too many deflections shall cause water logging on different parts of the roof, thereby resulting in corrosion on these locations, and hence need to be controlled. Since the roof layout is similar to that of the bottom plate, a similar welding sequence developed for bottom plate lap welding may be followed here also. Furthermore, care shall be taken not to overweld any of the joints in order to control the quantity of the weld metal on the roof plate and thereby the resultant buckling.

Once the roof is completely erected, the roof appurtenances are fitted as per the design. This shall be followed by platforms and other structural items that are to be erected on the roof, including lighting and lightning protection devices and other instrument connections as required in the design.

Roof construction: bull wheel and center column in progress.

Roof rafter construction in progress.

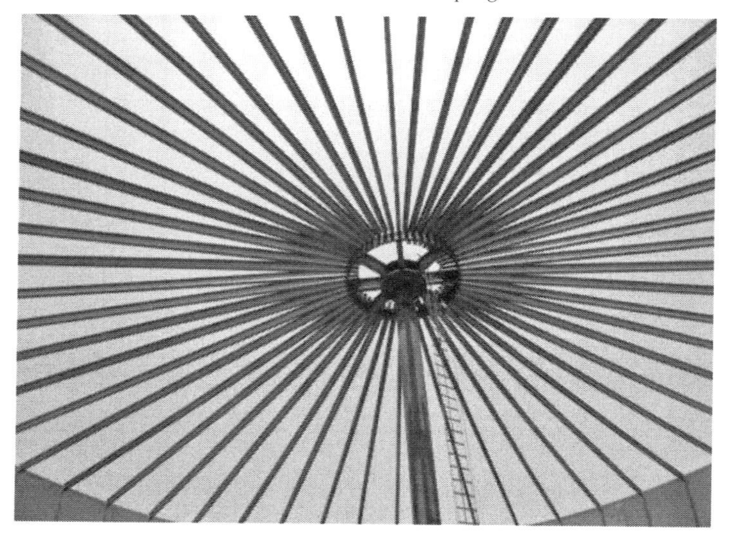

Supported roof partially completed.

Center column support (after painting).

8.5.4 Fit Up and Welding of Closure Plate

The last odd-size plate to be fitted to complete the bottom shell course is called the compensating or closure plate. This is fitted only upon completion of welding of all vertical seams of the first shell course except the last two vertical seams of this plate, and manhole or clean-out door openings are made on the shell. Usually the length of the closure plate shall be more than the width of the shell courses for easy movement of plates (if required) and other internals to be mounted inside the tank. Oil storage tanks are expected to have large internal distributors on the inlet side and also suction headers on the outlet side.

The compensating plate in each shell course, which is the last shell section of the respective shell courses, takes care of any resultant shortage in actual circumference of the rest of the shells upon completion of vertical seam welding. To have this flexibility, the closure plate is prepared only after completion of erection and welding of all other plates of that shell course. Even though the size of the closure plate is mentioned in the drawing, changes may have to be made at the site to suit site conditions to obtain the precise outside circumference as required and indicated in Table 8.1.

Once the circumference requirement is finalized for the closure plate, the same is marked and edges are prepared as desired in the drawing. The plate is then bent at the required radius and erected as the other shells were. Fit up and welding also are carried out in same way. Apart from trimming the length to suit site requirements, erection is the same as that of any other plate of that shell course.

8.5.5 Fit Up and Welding of First Shell Course to Annular Plate

As mentioned already, the first shell course is erected on the annular plate. The first step in this regard shall be marking the inside and outside circumferences of the first shell course on the annular plate as per the drawing. Wherever this marking crosses butt welds between segments of annular plates, reinforcement of butt welds is ground flush with the parent metal to facilitate proper placement and seating of the shell course on the completed annular plate. This flush grinding of the reinforcement is done for a length of about 25 mm on either side of the circumferences marked on the annular plate.

Shell plates are squared, and edges are prepared as required in the drawing. The sizes of all individual shell plates shall be provided in the shell cutting plan, in which all other plates except the closure plate shall be of full size. Since the orientation of the tank is already fixed before laying the annular or bottom plates, shell plates are erected according to the prefixed orientation of the bottom plate. Shell plates are then erected progressively from one end to the other excluding the closure plate, which is erected at the end.

Usually, the weld between the annular plate and the first shell course shall be a fillet weld, with or without a "V" on the shell plate. Some client specifications insist for a 45^0 "V" bevel of the shell plate at the joint. Shells are aligned vertically as per the shell cutting plan along the circumference already marked on the annular plate. Stiffeners are provided at fixed intervals to retain plates in a vertical position along the circumference. Since shell plates are already bent to the required diameter, this process is comparatively easy. Upon completion of erection of all shell plates, the circumference at the bottom and top of each shell course is taken. This shall be within acceptable limits as specified in the code or client specification, whichever is more stringent. The weld joint configuration of the shell to the annular plate joint and all vertical seams are then inspected for profile at the joint, "V" geometry, and uniformity in edge preparation throughout the entire joint (especially root face and gap). In case these are within acceptable limits as per applicable documents, the same can be released for welding.

8.6 Methodology for Tank Erection by the Jacking Up Method

8.6.1 Brief Overview

An outline of the methodology goes as follows. Though the erection of a specified shell course is the same as that in the conventional method described in earlier sections, in this method, instead of erecting the first shell course first, the last or topmost shell course is erected first.

The tank bottom plates are placed on the prepared foundation and welded together as described in Chapter 7. Spacers cum guide beams (or pipes) of not more than 400 mm height (stools) are tack welded to the tank bottom along the periphery of the shell plates, already marked on the annular plate. The shell circumference is again marked on the stool surface plate, for easy erection of the shell plates. The shell plates of the first shell course (the top ring of the tank) are positioned on these temporary chairs, aligned, fitted up using temporary clamps, and welded together.

Lifting equipment (jacks, trestles, and hydraulic pumps) is assembled as shown in Figure 8.2. Further shell plates of the second shell course (next to the top course) are positioned outside the first one. The completed part of the tank (top shell) is lifted hydraulically to a height so that plates of the next shell course can be moved into place immediately.

Plates of the second shell course are located exactly as per the drawing with proper staggering of the vertical seam, and all vertical seams are welded. The horizontal seam between courses is also welded in a progressive manner. This shall be followed by welding the attachments and accessories to this shell course, erecting the roof structures, and laying the roof plates.

Upon completion of the horizontal seam welding (between the top two shell courses) or while it is progressing, the roof structure installation can be taken up, followed by laying the roof plates.

While welding of the roof plates is being completed, all appurtenances and accessories such as handrails, platforms, supports on the roof, portions of the staircase, and similar accessories on the top two shell courses also shall be completed in all respects, including the NDT and all other inspections. However, one or two roof plates (based on the size of the tank) are not placed, usually to facilitate circulation of air during the entire construction operation. These plates are welded once all internal welding and assembly works are completed.

Upon completion of these steps, the plates of the third shell course are placed outside the second shell ring, and the portion of the tank is now ready for lifting. This cycle of operations is repeated until the last (bottom) shell ring is fitted and welded to the second shell course from the bottom of the tank. The entire tank is now lowered down to the annular plate. On completion of the annular plate to the shell welding, the lifting equipment is dismantled.

8.6.2 Detailed Working Procedure

- The tank periphery (shell ID and OD) is marked out on annular plates.
- Spacers cum guide beams for guiding shell courses are placed at approximately 1,500 mm–2,000 mm spacing, at a fixed level, and are tack welded to the annular or bottom plates.

- The first (uppermost) shell course is positioned on leveled spacer cum guide beams, and the level of shell ring also shall be checked as reconfirmation. Shell plates shall be placed exactly as per the contour (radius) marked on the guide beam surface plate and shall be truly vertical (within the specified tolerance).
- Welding, inspection, and NDT follow.
- The roof structure is assembled and fastened to the top shell course. Most of the roof plates are placed and welded (except for one or two) for easy air circulation.
- The location of lifting trestles is marked out inside the erected shell course at equal spacing.
- The lifting equipment (described in Section 8.6.4, Figure 8.2) is installed.
- The first shell course is lifted to one plate height.
- The second shell course (from top) is erected next.
- While the horizontal seam welding between these two shells is progressing, the roof structure assembly and welding followed by the laying of roof plates and its welding are carried out.
- The completed part of the tank is lifted so that the next shell ring can be inserted below the first two shells. Care shall be taken not to lift too much above the requirement.
- Lifting can be stopped at any moment by the operator by cutting fluid flow to the jacks. In practice, lift may not be the same at all points. As soon as the required height is reached at any one point, lifting is stopped, and the stop valve of that lifting point (jack) is closed. Lifting is continued until the required height is reached at all lifting points.
- All valves are to be closed, and the elevated section of the tank is now loaded on lower grip-jaws of the climbing jacks.
- Adjustment of the tank level (if necessary) can be made by lifting or lowering with one or more jacks at a time. While doing so, the valves on the other jacks shall be closed.
- The shell course is fitted and welded (vertical joints followed by horizontal joints). For easy movement of personnel and materials, the erection and welding of the last (closure) plate of each course is carried out as the last operation in each shell course erection.

- All welding works (including the welding of the closure plate) on respective shell courses shall be completed prior to lift up. This includes vertical and horizontal welds (from both sides) and that of all appurtenances, attachments, and accessories.
- Climbing jacks and sliding chairs are lowered, one at a time, using a rope with chain. During this operation, both of the grip-jaw sets of the jack shall be in a disengaged position.
- Lifting and fend-off lugs are now welded to the next uppermost shell course.
- This procedure is repeated for all following shell courses. As the shell courses are added, the load on the lifting trestles increases, which calls for higher oil pressures. However, the oil pressure shall not be increased exorbitantly, and in no case should this exceed 135 kg/cm^2.
- Before assembly of the last (bottom) shell course, all the spacers cum guide beams are removed. If necessary, a hydraulic power pack also can be placed outside the tank.
- After the last shell ring is welded to the tank wall, fit up and welding of the bottom shell to the annular plate joint is carried out.
- Lifting attachments and equipment are dismantled and removed through the last plate of the bottom shell course. This plate is welded as the last operation.

Note that lifting of the tank shall not be carried out when wind velocity exceeds "maximum safe wind speed for lifting," which can be obtained from the tank erection specialist. Weather forecasts shall be referred to in this regard to select proper timing for lifting.

8.6.3 Salient Features and Limitations of the Jacking Up Method

Salient Parameters in Jacking Up

Sl. No.	Parameter	Limitation
1	Plate width	Up to 3 m
2	Unitary climb	100 mm/stroke
3	Lifting capacity	12 T
4	Plate thickness	Up to 30 mm
5	Time per stroke	5 min
6	Working pressure Hydraulic system	200 kg/cm^2

8.6.4 Arrangements for Jacking Up

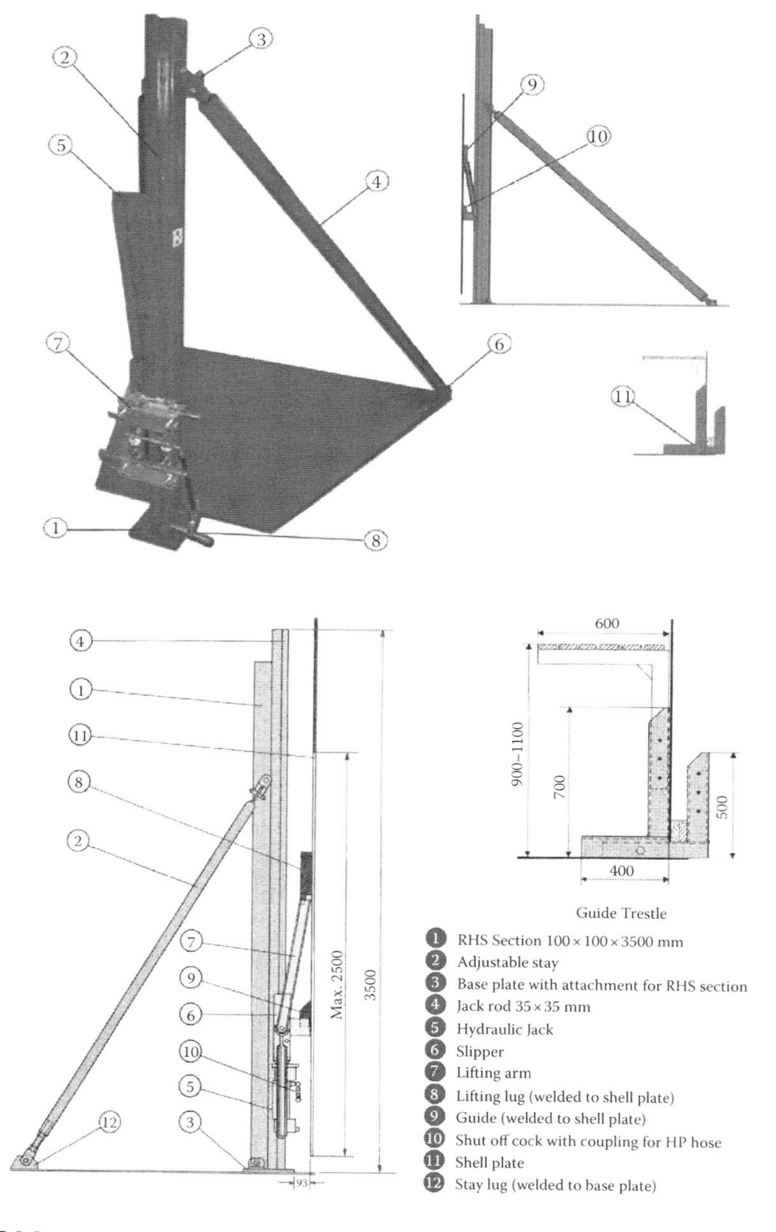

FIGURE 8.2
Main components of tank trestles. Drawing Courtesy M/s Bygging Uddemann AB.

8.6.5 Pictures of the Jacking Up Method

Jacking up arrangement (view from inside).

Note: Picture courtesy M/s Bygging Uddemann AB.

Tank in lifted position (inset hydraulic pump).

8.7 Erection of Roof Structure and Cone Roof Plates

In the jacking up method, the fit up and welding of the curb angle or compression ring is carried out after the completion of erection and welding of the two shell courses at the top. In the conventional method, this activity can be taken up only after erection of the top shell course, and all this activity shall take place at the full height of the tank. In both cases, this is followed by the erection of the fabricated center drum, roof truss, girders, and cross girders as per the drawing. The roof structure installation involves both welding of supports and bolting of the structure to the shell supports and center drum. The roof structure welding also shall be carried out by welders qualified to the approved welding procedure specification for the purpose.

Later roof plates are laid on the structure as per the drawing. During fit up of the short and long seam lap joints, proper lap as required in the drawing shall be maintained. To minimize distortion of the roof plate due to welding, it is recommended to follow sequential welding as in the case of the bottom plate. Since the layout plan of the roof plate also resembles that of the bottom plate, the welding sequence developed for the bottom plate as shown in Table 7.2 can be adopted here to minimize distortion.

Along with these works, roof nozzles, manways, and other attachments are fit up and welded. Ensure that nozzles, manways, and attachments are positioned exactly as required in the drawings.

8.8 Shell Appurtenances

In the conventional method of shell erection, nozzles, manways, and other attachments can be welded at any point during the time after the tank has acquired enough rigidity. In the jacking up method, shell appurtenances and attachments, including spiral stairways, have to be completed along with each shell course erection. Otherwise, scaffolding may have to be erected again for this purpose.

8.9 Manufacture of Subassemblies

Items that are categorized under this heading are those fittings and attachments that can be manufactured simultaneously with the construction of the tank as a parallel activity. These include, but are not limited to, the following:

- clean-out doors (COD)
- manways (shell and roof) and davits, if any

- nozzles
- anchor chairs
- curb angles, wind girders, compression ring, etc.
- pipe support cleats and supports
- roof support chairs
- platform and ladder supports
- distributor supports and chairs
- platforms, ladders, etc.

These items are manufactured separately according to priority well in advance. Nozzles and manways may require NDT like RT, which may be a bit time-consuming and hence to be borne in mind while planning. Similarly, clean-out doors often call for post weld heat treatment (PWHT) after manufacture and NDT, prior to installation. Since a PWHT facility is not required at the site, this often has to be outsourced locally. This also has to be considered while planning subassembly works.

Since manway sizes (as per code) range from 500 mm to 900 mm, manway necks have to be manufactured from the plate by bending. This also requires radiography for its long seam. Similarly, some client specifications require RT for nozzle necks to flange welds as well. Prior to the release of these subassemblies for installation, it shall be ensured that all required inspections and tests on these components are completed, so that the only remaining inspection shall be that related to installation and welding on the tank.

8.10 Installation of Appurtenances

The flange to pipe joint shall be prefabricated, and the required NDT shall be completed before fit up. Prior to this, orientation and elevation shall be marked on the shell and roof as per the drawing. Since this happens at elevation, manual gas cutting alone shall be possible in most circumstances. Highly skilled gas cutters shall be deployed for such works so that rework can be eliminated to the extent possible. Upon making a straight cut opening on the shell, proper edge preparation also shall be done by gas cutting or grinding as applicable. Fit up of prefabricated nozzles and manways is carried out as per the drawing at the required orientation and elevation with required standoff for nozzles. Large-size nozzles and manways shall require reinforcement pads as well and have to be installed along with nozzles or manways as the case may be.

Suitable jigs and fixtures shall be provided during welding to reduce distortion during welding. Moreover, to accommodate for shrinkages manifested as shortage in standoff during welding, all nozzles shall be provided with at least 3 mm to 5 mm excess projection during fit up. Manway necks and other large-size nozzle necks may require radiography of their welds, based on code or client specifications.

Fixed roof mushroom circulation vent.

Internal ladder with slotted gauge pole.

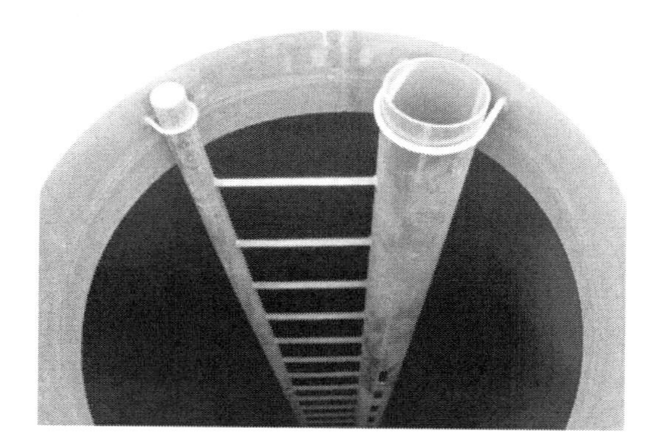

Ladder, slotted gauge pole, and sump.

Sample hatch.

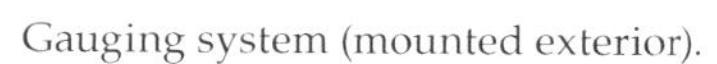

Gauging system (mounted exterior).

Inlet Diffuser

Suction Line Trough

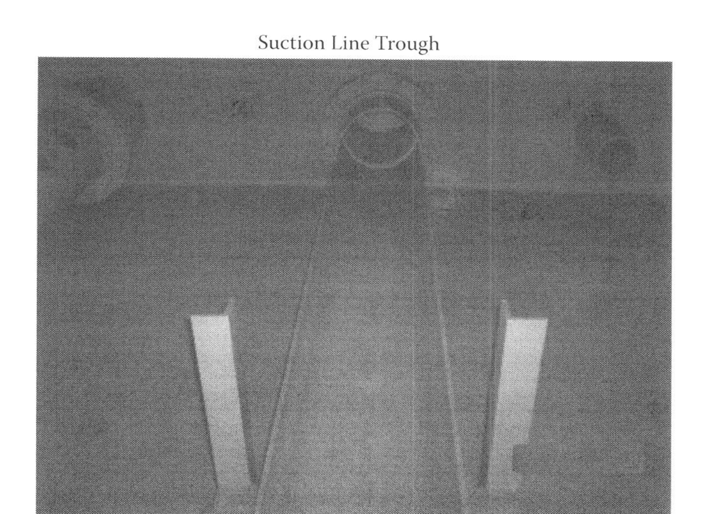

Water draw off piping.

Tank bottom sump.

8.11 Spiral Stairway, Handrails, Platforms, and Other Support Attachments

All structural items shall be straightened before marking and cutting to required dimensions. Gas cutting is usually adopted for this purpose. Irregular cut edges shall require grinding to remove irregularities and to remove oxidized material from the welding face. These are to be fabricated according to applicable drawings and specifications, and each of the items shall undergo at least visual inspection prior to surface preparation and protective coating.

Dimensions, weld size, appearance of welds, and so on, shall be critera of inspection. Removal of slag, presence of undercuts, and other surface defects on the welds also shall be given due consideration.

Spiral stairway.

9

Welding

9.1 Weld Edge Preparation

Edge preparation for any pressure and nonpressure joints shall be in accordance with the design given in the concerned drawing prepared according to code and specifications. Variations in the configuration (such as a change in "V" angle, etc.) shall be within normal allowable tolerances. Variation of root gap is one of the major problems seen during fit up of groove welds. Variation of root gap shall be limited within 1 mm to 1.5 mm throughout the entire weld joint (for vertical seams). Though this is not specified anywhere in API 650, provided proper care is taken during edge preparation and subsequent handling and fit up of the shell plates, this tolerance is quite possible. If this is achieved, its impact shall be seen in the final welding, and chances of having a repair shall be much less compared to a joint with more variation in root gap.

Another prominent defect usually observed is variation in root face or throat thickness. This may give rise to weld defects such as lack of fusion or insufficient fusion. With proper care taken during edge preparation, this can be reduced to a minimum. The preferred root face for shielded metal arc welding (SMAW) is in the range of 0.9 to 1.2 mm, which could be duly established while qualifying the welding procedure specification (WPS) (as this is principally dependent upon the current used and the manipulation of the welding rod in root welding).

9.2 Typical Weld Joints

The various types of weld joints that are considered acceptable as per API 650 are provided in the following table, with reference to the applicable sketches in the API 650 2013 edition.

Though the standard specifies various options in selecting weld joint details, its judicious selection depending on the situation is the prerogative of the manufacturer, which again is open for comments from third-party inspection agencies, consultants, clients, and statutory authorities.

9.3 Restrictions on Types of Weld Joints

API 650 specifies certain general requirements (irrespective of materials) with regard to the joint preparations referred to in Table 9.1. The restrictions on the types of weld joints are consolidated in Table 9.2. As this table provides only a gist, to understand the finer details of these restrictions, refer to Section 5 of API 650.

TABLE 9.1

Weld Joint Guide

Type of Joint	Reference Sketch in API 650
Annular, Sketch, and Bottom Plate Weld Joints	
Single welded full fillet lap joint	Figure 5.3a
Single welded butt joint with backing strip and optional groove	
Typical Weld Joints for Vertical Seams in Shell	
Square groove butt	Figure 5.1
Single "V" butt	
Single "U" butt	
Double "V" butt	
Double "U" butt	
Typical Weld Joints for Horizontal Seams in Shell	
Square groove butt (FP)	Figure 5.2
Single bevel butt (FP)	
Double bevel butt (FP)	
Angle to shell butt (FP)	
Alternative, angle to shell	
Typical Roof Plate and Roof to Curb Angle Joints	
Roof plate joint	Figure 5.3a
Roof to shell joint	
Alternative, roof to shell joint	
Typical Joints (Shell to Annular Plate)	
Annular plate below 13 mm	Figure 5.3a
Annular plate equal to or above 13 mm	Figure 5.3c
Typical Lap Preparation under Shell	
For all thickness	Figure 5.3b

Note: FP, full penetration.

TABLE 9.2

Restrictions on Type of Weld Joint

Sl. No.	Type of Weld	Restrictions
1	Vertical shell joints	Full penetration and complete fusion are required. Vertical joints are to be offset by a distance of 5 T (where T is the thickness of the thicker of the two).
2	Horizontal shell joints	Full penetration and complete fusion are required. For top curb angles, double welded lap welding is permitted.
3	Lap welded bottom joints	Plate edges shall be free of deposits. Laps are in the direction of the slope. Three plate laps shall be 300 mm away from each other, the tank shell, and the butt weld of the annular plates and from the joint of the annular plate to the bottom plate. Lapping of two bottom plates onto butt welded annular plates does not constitute a three plate lap weld. Bottom plates need to be welded on the topside only, with a continuous full fillet weld on all seams.
4	Butt welded bottom joints	Square or "V" grooves, full penetration, and a backing strip of a minimum of 3 mm thick are permitted. For a square face, the minimum gap is 6 mm. Three plate joints shall be at least 300 mm from each other and from the tank shell.
5	Bottom annular plate joints	Complete penetration and fusion are required. A backing strip is permitted.
6	Shell-to-bottom fillet welds	If the bottom and annular plates have a nominal thickness of 13 mm or less, use a continuous fillet weld on either side. For thickness above 13 mm, a fillet or groove weld is permitted.
7	Wind girder joints	Full-penetration butt welds are required. Continuous welds are required for the topside. For the bottom side, it shall be a seal weld.
8	Roof and top-angle joints	A continuous full fillet weld on all seams is required on the top surface. Butt welds are also permitted. Top-angle sections, tension rings, and compression rings shall be joined by butt welds having complete penetration and fusion.
9	Fillet welds	A full-size fillet weld is required up to a plate thickness of 5 mm. Above 5 mm, 1/3 t is required, where t is the thickness of the thinner plate, with a minimum size of 5 mm.

Though Tables 9.1 & 9.2 indicate typical acceptable joint configurations with restrictions permitted by code, as said earlier, the judicious use of joint welds from an economic, weld distortion, and quality point of view is definitely the prerogative of the tank manufacturer. Apart from the selection of the weld joint, the selection of the welding sequence also plays a vital role in controlling distortion in weld joints. From this point of view, the following methodology is suggested for tank shell welding.

9.3.1 Shell Vertical Joints

In the jacking up method, the shell is erected from top to bottom. In the conventional method, the shell is built the other way around. In both cases, the erection methodology followed is the same. Assembly of the first shell course (either top or bottom) is done in position, based on the circumference marked on the annular plate. In the jacking up method, shell plates are erected on spacer stools, whereas in the conventional method it is erected directly on an annular plate. Plates are set up one after another, and welding is normally carried out after the release of the entire shell course for vertical seam welding, except the two seams on either side of the closure plate.

Upon erection of all plates of a shell course except the closure plate of that particular shell course, more welders are usually deployed simultaneously all around the circumference with a safe distance between each welder in order to meet stringent time schedules associated with the project.

For a tank 38 m in diameter, the approximate outside circumference is around 119.3 m. Assuming a plate length of 12 m, the entire circumference can be formed using nine full-size plates and a compensating plate of about 11.3 m length. This constitutes ten vertical weld joints. For easy movement of workers, materials, and machinery to and from the tank, the compensating (closure) plate shall be left open till completion of all works, and this plate is welded at the last stage (for conventional erection). Therefore, the number of vertical seams to be welded in the first phase is eight. If alternate vertical seams are attempted simultaneously, this will provide a distance of around 24 m over circumference between welders. This requires four qualified welders at a minimum. If more welders are available, proper sequencing can be done at the site. The usual shell plate width is around 2.5 m to 3 m, and one welder can easily handle one seam. Such an assignment will exercise better control over quality and hence is recommended.

When welding is being done from the inside as well, the same number of welders can be deployed. From the distortion point of view, it is better if seams are welded simultaneously from the inside and outside. This requires deployment of additional welders. If this is done, proper time lag needs to be maintained between the start of the outside and inside welds. This is not possible for root and hot passes of groove welds in a manual process, which would require back gouging (in manual processes, the second side weld can be initiated only after back gouging and grinding of the second side to ensure sound base metal at the weld joint). In the event of double "V" preparation (for a shell thickness greater than 16 mm), the major "V" is usually provided on the outside, requiring more passes from the outside wherein more welders can be deployed comfortably, as they are in the open.

As far as possible, the same welder shall complete the weld, though it is permitted to have multiple welders for one seam as per code. If there is simultaneous welding from the inside and outside, this is not practical. Since the quantum of welding from the outside is more, more welders are required for the outside compared to the inside. As an option of manufacture, welding from the inside can be started either after completing welding from the outside or after depositing two or three passes from the outside.

The last two joints are welded only after the closure plate is fitted. The closure plate is usually fitted only after all works are completed on the shell course from both the inside and the outside. As mentioned earlier, in the jacking up method, tank erection starts from top to bottom, after bottom plate erection and welding. In such cases, work on the top-most shell includes the erection of roof plates and other structural members including ladders, platforms, supports, and so on. Completion of these works is absolutely essential, since carrying out these works at height poses safety hazards and also requires scaffolding again after erection.

Upon completion of all works on the shell course, the compensating plate or closure plate is fitted. The dimensions of this plate are decided based on the actual conditions at the site. The welding of two seams on the closure plate is also carried out as recommended for other vertical joints. In the jacking up method, each shell course shall be allowed to lift only after the erection of all shell plates that are completed and fully welded. Jacking up the shell with partial welding of the shell courses shall not be permitted under any circumstances. This also requires completion of welding from the inside and outside of closure joints prior to lifting. As the shell plate is set up on stools of about 400 mm to 500 mm height, entry of workers into and out of the tank shall still be possible. The criteria for fixing stool height will be decided based on the width and height of the shell plate or sections to be used for construction of the shell and roof structures. If the width of the plate is more than 2,500 mm, it is always better to limit the stool height to about 400 mm to 450 mm so that entry of workers will still be possible. Excessive stool height may lead to the requirement of a platform to complete vertical seam welding of every shell course, especially when wider plates are used.

On completion of each shell course, it shall be lifted up, and the second course from the top shall be erected as per the staggering proposed in the drawing so as to avoid fouling of nozzles and other attachments with the main weld seams. In the conventional method, the difference is that shells are erected from bottom to top, and the closure plate on the first shell course is usually welded as a last operation, in addition to clean-out doors and manways for easy movement of workers, materials, and machinery.

9.4 Welding Processes

Usual welding processes that are permitted by API 650 (Clause 7.2.1.1) are SMAW, gas metal arc welding (GMAW), gas tunsten arc welding (GTAW), oxy fuel gas welding (OFW), flux cored arc welding (FCAW), submerged arc welding (SAW), electro slag welding (ESW), and electro gas welding (EGW) using appropriate equipment for the same. However, use of OFW, ESW, and EGW shall be employed only if the purchaser agrees for the same.

9.5 Welding Procedure Specifications

Upon receipt of the order for storage tanks, based on preliminary fabrication drawings, the quality assurance team of the manufacturer prepares a comprehensive list of welding procedures required, including that for structural steel welds on the tank. For tanks constructed as per API 650/620, WPS and its qualification shall be carried out as per ASME Section IX. WPS shall be capable of addressing all expected combinations of weld sizes, base materials, and thickness ranges as required in respective drawings. This shall include all structural steel attachments to the tank shell as well. However, for ladder and platform assemblies, handrails, stairways, and other miscellaneous subassemblies, WPS need be qualified, either to AWS D1.1, AWS D1.6, or Section IX of ASME code, including the use of standard WPS.

In the case of established tank erectors, many (if not all) required WPS might be qualified already, and records may be available for review. In such circumstances, WPS need not be requalified. However, the decision of the inspection agency or client shall be final and binding in accordance with the contract. Though it is not mandatory that all WPS need to be qualified under third-party inspection, in the case of procedures already qualified by tank erectors, many clients require a third-party witness for the procedure qualification record (PQR) to improve its authenticity and as a minimum requirement for consideration.

9.6 Procedure Qualification

In the event of new WPS, a PQR test shall be carried out at the site or factory, well in advance based on a proposed WPS, so that all mechanical tests required for qualification as per ASME Section IX can be completed prior to the start of work at the site. Obtaining approval of all necessary WPS with their PQRs shall be the responsibility of the QA team of the manufacturer (either at its works or at the site), and a controlled copy of this document shall be available with the site QC crew to proceed with the qualification of welders and welding operators as required at the site.

9.7 Welder Qualification

Once the PQR is qualified, the next important activity is the qualification of welders or welding operators. Since the construction of storage tanks is predominantly a site activity, welders are normally recruited at the site. Because of this, welder and welding operator qualification tests are usually carried out at the site. The number of welders to be qualified against each process shall be judiciously decided based on the estimated quantum of welds in each category and time schedule for the entire construction activity according to the work plan.

Depending on production targets, available welders are qualified under different WPS. It would be advantageous to have a few more welders qualified than required so as to cater to voids created by unexpected migration of welders between contractors within the region. Welder or welding operator qualification also shall be carried out as per ASME Section IX, and certificates to this effect shall be issued at the site, the original of which shall be signed by all concerned agencies and shall be available at the site. Furthermore, it is desirable to issue the qualified welders with welder's identification cards with photos so that unauthorized welding can be detected easily during surprise surveillance checks by inspecting agencies.

Welder qualification testing shall be carried out strictly in accordance with duly qualified WPS, and the welded test specimens shall be subjected to either a bend test or radiographic testing (RT) as required in ASME Section IX.

Though not mentioned in ASME Section IX, some client specifications allow welder qualification tests to be terminated at the welder's option if he becomes aware that he has introduced a defect, and may be restarted after re-preparation or replacement of the test specimen. Only one such welder's optional restart may be allowed except for intended restarts for new electrodes or repositioning.

On completion of the test weld, no repairs shall be allowed, including defects that are revealed by cosmetic grinding or filing as permitted.

The contractor, consultant, or client may terminate the test whenever it becomes apparent that the welder lacks the required skill to produce satisfactory results, or if he or she fails to observe the requirements of WPS or observe safety norms.

In the event of a test failure, the consultant or client reserves the right to disallow an immediate retest and to insist on a period of further training before any retest.

While ASME Section IX test requirements are sufficient for all commonly used welding processes like SMAW, SAW, and so on, additional test requirements are specified by clients for welder or welding operator qualification for processes like SAW or EGW. The additional test usually recommended is ultrasonic examination of the weld in addition to either bend testing or radiography as required in ASME Section IX for qualification of welders and welding operators.

9.8 Welder's Identification Cards

Welders shall be assigned an identification code prior to undertaking any tests. A unique test number shall be allocated to each test and stamped on the specimen.

All test documentation shall reference this test number. Identification, grade, and thickness of the parent test material shall be positively established and recorded.

On satisfactory completion and recording of the welder's test, a card shall be issued to each welder with the following information as a minimum:

- Photograph of the welder (not facsimile)
- Welder's name
- Welder's identity code
- Brief details of qualified processes, positions, thicknesses, diameters, and consumable codes
- Consultant or client approval stamp, signature, and date
- Validity of card

Cards with at least the welder's photograph, name, and identification number shall be presented at the time of testing as a means of positive identification. Welders shall be instructed to carry their identification card at all times when welding work is being performed.

9.9 Welding Sequence

Irrespective of the welding process or joint configuration, the following methodology is recommended for carrying out any weld.

9.9.1 Cleanliness of Weld Groove and Adjacent Area

Cleanliness of the weld groove plays a vital role in the final quality of weld, often revealed through nondestructive testing (NDT). With a little effort or precaution, this can be ensured, which adds significantly to the quality of the weld. After fit up (of any type of joint, butt, or fillet), just before the start of welding, groove or weld deposition surfaces shall be brush cleaned with a power brush so as to remove any rust or dirt that has accumulated because of humidity in the atmosphere or collection of dust from the atmosphere. Prior to the start of the weld from the groove side (for single bevel grooves), the joint shall be cleaned from the rear side as well. In addition to cleaning the groove, cleaning of about 10 mm width on either side of the groove would reduce chances of any debris going into the weld metal, which might affect the quality of the weld. Though this requirement is usually specified for groove welds, this may be extended to fillet welds to improve their quality.

Since tank construction is purely a site activity, atmospheric conditions also play a vital role in the final weld quality. Under no circumstances shall welding be started on a wet surface or during rainfall or icefall or dust storms. Furthermore, strong winds at the site also could be a reason for weld defects. So, providing adequate protection from wind and other adverse atmospheric conditions (proven to be detrimental to the weld) is absolutely essential from a quality point of view.

9.9.2 Metal Temperature and Preheat

Because tank construction is a site activity, mostly in the open, yet another important factor is the metal temperature. While this is not a significant issue for tropical countries, in places where the temperature falls as low as 10°C or below, this is a critical situation. In such instances, after the weld groove and adjacent area are cleaned, the metal temperature needs to be checked with an infrared thermometer.

As per API 650, the lowest temperature permitted without preheat for any material is 0°C (32°F), whereas ASME Section VIII Div (1) Appendix R specifies 10°C (50°F) as the minimum temperature for welding in carbon steels without preheat. The author is of the opinion that 10°C (50°F) is a more rational limit for this lowest temperature. Table 7.1a of API 650 provides a minimum preheat requirement for welding of different materials covered by the standard. As per this table, for most commonly used carbon steel materials, the temperature ranges from 0°C (32°F) to 93°C (200°F) based on the thickness of the component to be welded.

At ambient temperatures below 10°C (50°F), additional measures shall be taken, such as shielding components, doing extensive preliminary heating, and preheating. This is significant especially when welding with a relatively low heat input (energy input per unit length of weld), for example, when laying down thin fillet welds, where rapid heat dissipation occurs, or when welding thick-walled components. Wherever possible, it is recommended not to carry out welding at ambient temperatures below –10°C (14°F).

Preheat is commonly applied with fuel gas torches or electrical resistance heaters and has to be applied over a band of at least 75 mm width on either side of the groove or of the weld area.

9.9.3 Reasons for Preheating before Welding

The need for preheating ferritic steels and the preheating temperature depend on a number of factors. The predominant among them are as follows:

- Chemical composition of the base material (carbon equivalent) and the weld metal
- Thickness of the work piece and the type of weld joint (two- or three-dimensional heat flow)
- Welding process and the welding parameters (energy input per unit length of weld)
- Shrinkage and transformation stresses
- Temperature dependence of the mechanical properties of the weld metal and the heat-affected zone
- Diffusible hydrogen content of the weld metal

9.9.4 Interpass Temperature

Interpass temperature refers to the temperature at the weld just prior to the deposit of a new weld layer or pass. It is identical to preheat, except that preheating is performed prior to any welding. When a minimum interpass temperature is specified in WPS, welding shall not be performed when the base plate containing already-deposited weld is below this temperature. The weld shall be heated back to the required minimum before welding continues. For certain other materials, the maximum interpass temperature may be specified to prevent deterioration of the weld metal and heat-affected zone properties. In this case, the weld shall be below this temperature before welding continues.

API 650 is rather silent on this matter. However, a general guideline is provided in API 582 as follows based on the P. No. classification in ASME Section IX.

Material Group		
P. Number	Description	Minimum Interpass Temperature
P-1	Carbon Steels	315 °C (600 °F)
P-3, P-4, P-5A, P-5B, and P-5C	Low Alloy Steels	315 °C (600 °F)
P-6	Type 410	315 °C (600 °F)
P-6	CA6NM	345 °C (650 °F)
P-7	Type 405/410S	260 °C (500 °F)
P-8	Austenitic Stainless Steels	175 °C (350 °F)
P-10H	Duplex Stainless	150 °C (300 °F)
P-41, P-42		150 °C (300 °F)
P-43, P-44, and P-45		175 °C (350 °F)

9.9.5 Back Gouging, Grinding, and NDT

Back gouging is the removal of weld metal and base metal from the root side of a welded joint to facilitate complete fusion and complete joint penetration. This may be carried out either after full completion of welding or with partial welding on the major side as considered apt for the situation. Processes like carbon arc gouging or plasma may be used to remove material, followed by grinding to remove oxidized material from the rear weld groove to reach sound metal. The resultant groove on the hind side shall have adequate and uniform depth to ensure complete penetration into the previously deposited weld metal from the major side of the bevel.

To ensure that all root defects have been removed, it is even advisable to carry out dye penetrant testing on the hind-side groove produced through back gouging and grinding. When this is carried out satisfactorily, welding from the hind side can be started after bringing it back to the desired interpass temperature.

Upon completion of each weld in this manner, it shall be subjected to visual inspection. Visual inspection of completed welds may call for some cosmetic grinding, after which the weld can be released for NDT as applicable. A weld is considered acceptable when NDT (random spot or full as applicable) is completed satisfactorily on the weld.

9.10 Electrode Storage and Its Drying

Based on WPS, and considering the quantum of welds in each category of welds, proper welding electrodes (in case of SMAW) and flux and wire (for SAW) and applicable consumables (for other welding processes) are estimated and ordered with staggered delivery depending on the construction schedule. Often orders are placed with well-reputed manufacturers of welding consumables. Once these materials start arriving at the site, storage-related issues arise and need to be tackled.

Electrodes for SMAW or stick electrodes must be properly stored in order to deposit quality welds. Stick electrodes absorb moisture from the atmosphere, and they need to be dried in order to restore their ability to deposit quality weld. Electrodes with high moisture content may lead to cracking or porosity or may affect strength characteristics of the deposited weld metal. In cases where unexplained weld cracking problems or problems related to arc stability are observed, it could be due to improper storage or the redrying procedures adopted.

The following storage and redrying tips might be helpful in deriving the highest quality welds from stick electrodes.

9.10.1 Storing Low Hydrogen Stick Electrodes

- Low hydrogen stick electrodes must be dry to perform properly. Unopened hermetically sealed containers provide excellent protection in good storage conditions. Opened cans should be stored in a cabinet at 120°C to 150°C (250°F to 300°F).
- Low hydrogen stick electrode coatings that have picked up moisture may result in hydrogen-induced cracking, particularly in steels with a yield strength of 550 MPa (80,000 psi) and higher.
- Moisture-resistant electrodes with an "R" suffix in their AWS classification have a highly resistant moisture pickup coating and, if properly stored, will be less susceptible to this problem, regardless of the yield strength of the steel being welded.
- All low hydrogen stick electrodes shall be stored properly, even those with an "R" suffix. Standard EXX18 electrodes should be supplied to welders twice per shift. Moisture-resistant types may be exposed for up to 9 hours.

When containers are punctured or opened, low hydrogen electrodes may pick up moisture, which may affect the final quality of the weld in the following ways:

- High moisture in the flux coating of low hydrogen electrodes may cause porosity, revealed mainly through radiography. For base materials and welds with higher strength, moisture can pave the way for underbead or weld cracking.
- A high amount of moisture also causes visible external porosity in addition to internal porosity. It can also cause excessive slag fluidity, a rough weld surface, difficult slag removal, and cracking.

9.10.2 Redrying: Low Hydrogen Stick Electrodes

When electrodes are exposed to humidity, redrying them prior to use is permitted if it is carried out properly as recommended by the manufacturer. The drying temperature and duration of drying required depends on the type of electrode and its condition.

- One hour at the recommended final temperature is satisfactory. *Do not* dry electrodes at higher temperatures. Similarly, drying at lower temperatures for extended hours is also not equivalent to using the specified temperature and time.
- Electrodes of E8018 and higher strength classifications shall not be subjected to more than three 1-hour redry cycles in the 370°C to

430°C (700°F to 800°F) range. This minimizes the possibility of oxidation of alloys in the coating resulting in lower than normal tensile or impact properties.

- Low hydrogen electrodes shall be discarded if excessive redrying has taken place, as this causes the flux coating to become fragile and flake or break off while welding. Electrodes shall be discarded if they show strange arc characteristics such as insufficient arc force or an unstable arc.
- Electrodes requiring redrying shall be spread out in the oven to receive uniform drying temperature.

9.10.3 Redrying Conditions: Low Hydrogen Stick Electrodes

		Final Redrying Temperature	
Condition	**Predrying Temperature**[a]	**E7018, E7028**	**E8018, E9018, E10018, E11018**
Electrodes that have been exposed to air for less than 1 week and had no direct contact with water	N/A	340°C to 400°C (650°F to 750°F)	370°C to 430°C (700°F to 800°F)
Electrodes that have come in direct contact with water or have been exposed to high humidity	80°C to 105°C (180°F to 220°F)	340°C to 400°C (650°F to 750°F)	370°C to 430°C (700°F to 800°F)

[a] Predry for 1 to 2 hours. This will minimize the tendency for coating cracks or oxidation of alloys in coating.

9.10.4 Storing and Redrying: Non-Low Hydrogen Electrodes

Electrodes in unopened cans or cartons retain proper moisture content indefinitely when stored in good condition.

By any chance, if electrodes are exposed to humid air for a long period of time, electrodes from opened containers may pick up enough moisture to affect operating characteristics and weld quality. In case moisture appears to be a problem, store electrodes from opened containers in heated cabinets at 40°C to 50°C (100°F to 120°F). *Do not* use higher temperatures, particularly for electrodes from the "Fast Freeze" group.

Some electrodes from wet containers or long exposure to high humidity can be redried. Adhere to the guidelines in the following table for each type.

9.10.5 Redrying Conditions: Non-Low Hydrogen Stick Electrodes

Stick Electrode	Electrode Group	Final Redrying Temperature	Time
E6010, E6011, E7010-A1,[a] E7010-G,[a] E8010-G,[a] E9010-G[a]	Fast Freeze: Excessive moisture is indicated by a noisy arc and high spatter, rusty core wire at the holder end, and objectionable coating blisters while welding. *Rebaking of this group of stick electrodes is not recommended.*	Not recommended	N/A
E7024, E6027	Fast Fill: Excessive moisture is indicated by a noisy or "digging" arc, high spatter, tight slag, and undercut. Predry unusually damp electrodes for 30–45 minutes at 90°C–110°C (200°F–230°F) before final drying to minimize cracking of the coating.	200°C to 260°C (400°F to 500°F)	30–45 minutes
E6012, E6013, E7014, E6022	Fill Freeze: Excessive moisture is indicated by a noisy or "digging" arc, high spatter, tight slag, and undercut. Predry unusually damp electrodes for 30–45 minutes at 90°C–110°C (200°F–230°F) before final drying to minimize cracking of the coating.	150°C to 180°C (300°F to 350°F)	20–30 minutes

[a] Predry for 1 to 2 hours. This will minimize the tendency for coating cracks or oxidation of the alloys in the coating.

9.11 Weld Repairs

Repair of weld joints is a very common phenomenon in the manufacture of storage tanks, and the main cause of this could be adverse site conditions or lack of skill. Apart from surface defects (visible to the eyes), internal defects are revealed only through nondestructive tests like RT, ultrasonic testing, and magnetic particle testing carried out after completion of welding and visual examination. Such defects have to be removed first and then rewelded using a qualified procedure. This procedure shall be available at the site, and personnel who are deployed at the site shall be well versed with contents of this document.

Based on the interpretation of radiographs, the defect location shall be marked on a weld seam, or a tracing of the same shall be provided to production in charge for the removal of the defect. It is recommended to remove the defect by grinding so that a person attempting the removal can actually see the defect, and removal can be ensured. Ensure that the resultant repair groove has a proper "V" angle for slag to float up, so also to provide enough width and length at the groove for the welder to manipulate the electrode and properly fill up the repair groove. As far as possible, try to carry out welding in a down-hand position, as this is the most favorable as far as weld quality is concerned.

9.12 Weld Repair Procedure

Every manufacturer shall have a repair procedure and normal procedure applicable to the welding process under consideration. In case repairs occur in automated welding, it may not be possible for local repairs (for very small lengths) to be undertaken by this process again. In such cases, the repair is carried out by manual welding. These requirements shall be reflected in the weld repair procedure, and the original applicable welding procedure shall be made as an annexure to this procedure.

Though it is not possible to anticipate all possible kinds of defects and their repair, the repair procedure need only address the removal of weld defects and the rewelding of groove welds under different processes envisaged in the construction process.

10

Inspection and Testing of Welds

API 650 code provides the minimum general requirements for inspection against each type of weld envisaged in tank construction. However, depending on specific applications, clients usually add on further requirements through their specifications and data sheets prepared for contractors. The following table tries to consolidate inspection and test requirements applicable to code and the usual supplementary requirements enforced by clients in the oil and gas industry.

10.1 Summary of Inspection and Tests

The following summary provides an overview of various requirements. For finer details of the same, refer to Appendix T of API 650 and thereafter to relevant clauses of code as referred therein.

TABLE 10.1

Summary of Inspection and Tests

Sl.		Type of Testing									
No.	Description of Weld	RT	UT	MPT	LPT	VT	VBT	AT	FT	HT	Remarks
Annular/Bottom Plate											
1	Annular plate (butt)	X[2]			X[3]	X[1]	X[1]				a
2	Annular plate to sketch and bottom plate					X[1]	X[1]				
3	Three plate lap welds				X[2]	X[1]	X[1]				
4	Draw off sump butt	X[2]			X[2]	X[1]					
5	Draw off sump fillet				X[2]	X[1]	X[1]				
6	Annular plate to shell full penetration/fillet			X[4]	X[3,4]	X[1]	X[3,4]				m
Shell											
7	Shell to shell vertical	X[1]				X[1]					b, q
8	Shell to shell horizontal	X[1]				X[1]					b, r
9	"T" joints between shell vertical/horizontal	X[1]				X[1]					b
10	Shell to compression ring				X[2]	X[1]					
11	Shell to curb angle	X[2]				X[1]					c, j
12	Wind girder to shell					X[1]					

(Continued)

Sl. No.	Description of Weld	Type of Testing									Remarks
		RT	UT	MPT	LPT	VT	VBT	AT	FT	HT	
Roof											
13	Roof plate fillet				X^2	X^4	X^1	X^4			d
14	Curb angle to roof plate					X^4	X^1	X^4			d
Shell Nozzles/Manholes											
15	Nozzle/manway butt (L seams and first C seam closest to shell)	X^1				X^1					e
16	Nozzle/manway butt (flange to neck)	X^2				X^1					
17	Fillet (flange to neck)				X^2	X^1					
18	Neck to shell (inside)			X^4	X^4	X^1		X^1			p
19	Neck to pad and pad to shell			X^4	X^4	X^1		X^1			o
Clean-out Door											
20	Butt welds (L seams and first C seam closest to shell)	X^1				X^1					
21	Neck to flange				X^2	X^1					
22	Door neck to shell			X^1		X^1					
23	Door neck to pad			X^1		X^1					
24	Pad to shell			X^1		X^1					
25	Bottom reinforcement plate to transition plate weld			X^1		X^1					
26	Bottom reinforcement plate to annular plate			X^1		X^1					
27	Bottom transition plate to annular/bottom plates			X^1		X^1	X^1				
28	Shell/shell reinforcement pad to bottom reinforcement plate and transition plate			X^1		X^1	X^1				s
Roof Nozzles/Manways											
29	Roof to pipe					X^1	X^4	X^4			
30	Pipe to pad					X^1	X^4	X^4			
31	Pad to roof					X^1	X^4	X^4			i
Important Structural Welds											
32	Center drum weld					X^1					
33	Structural weld butt with splicing					X^1					
34	Structural fillet welds					X^1					
35	Bolting of structures					X^1					
36	Structural attachment welds (inside)					X^1					
37	Structural attachment welds (outside)					X^1					

Sl. No.	Description of Weld	Type of Testing									
		RT	UT	MPT	LPT	VT	VBT	AT	FT	HT	Remarks
Floating Roof Components											
38	External/internal floating roofs					X^1			X^1		
39	External floating roof drain pipe	X^2			X^2	X^1			X^1	X^1	t
40	Pontoon butt welds	X^2			X^1	X^1			X^1		g
41	Pontoon fillet welds				X^1	X^1			X^1		h
42	Pontoon corner welds				X^1	X^1			X^1		h
Others											
43	All attachment welds and weld removed areas			X^4	X^4						f, k
44	Completed welds on stress relieved components			X^4	X^4	X^1					f, l
45	Nonstructural attachments such as insulation clips (welded)			X^4	X^4	X^1					f, k
46	Nonstructural attachments such as insulation clips (capacitor discharge welded)					X^1					
47	Repair locations revealed in RT	X^1	X^2								n

Note: RT = radiographic testing, UT = ultrasonic testing, MPT = magnetic particle testing, LPT = liquid penetrant testing, VT = visual testing, VBT = vacuum box testing, AT = air leak test, FT = floatation test, HT = hydrostatic testing. MPT/LPT recommended by code after root pass and at every 13 mm weld deposit is superseded by MPT/LPT after full welding and hence not considered relevant, especially from an effectiveness point of view in practice. The penetrating oil test specified in code is replaced with LPT, especially for floating deck welds, as it is found more convenient and effective.

[a] Code requires only random RT (10% to 50%) with conditions as in Clause 8.1.2.9 of API 650, even otherwise recommended.

[b] Twenty-five percent shall be at tee joints.

[c] Recommended RT is applicable only for butt weld curb angles.

[d] Either VBT or AT is applicable for air tight roofs.

[e] Longitudinal seam is applicable for large-size nozzles/manways.

[f] Either MPT or LPT is feasible.

[g] LPT needs to be conducted on the outside surface because of the confined space inside the pontoon.

[h] For all welds exposed to the medium.

[i] Air test at 100 kPa (15 psi).

[j] HT is required if butt welded.

[k] Applicable to only material Groups IV, IV-A, V, and VI as per API 650.

[l] To be done before hydrostatic test.

[m] Inside and outside. For other options, refer to Clause 7.2.4.1 of API 650.

[n] UT is recommended to ascertain repair methodology (whether to excavate from the inside or outside) before repair and re-radiography.

(Continued)

Sl. No.	Description of Weld	Type of Testing RT	UT	MPT	LPT	VT	VBT	AT	FT	HT	Remarks

[o] Applicable if there are nozzles with reinforcement pads.
[p] Air test at 100 kPa (15 psi) is applicable if pad is provided.
[q] Refer to Clause 8.1.2.2 of API 650 regarding the finer details of the extent of RT.
[r] Refer to Clause 8.1.2.3 of API 650 regarding the finer details of the extent of RT.
[s] Applicable for both inside and outside.
[t] Primary drain water test at 350 kPa (50 psi). Drain valve is to be kept open during floatation test.
[1] Required as per standard.
[2] Usually required by client.
[3] Desirable from construction point.
[4] Optional between the two or three NDT method proposed.

10.2 Butt Welds

All welds require visual inspection to ascertain contour, reinforcement, undercut, and other surface opening defects on the weld. Therefore, this needs to be carried out from both sides. Visual inspection of welds is a powerful tool to ensure the quality of the weld, and visual inspection supplemented by DPT (in specific cases) shall be used to eliminate the presence of surface opening defects. Whatever is the contour of the weld (even when weld reinforcement is within acceptable limits), it shall be free from any abrupt changes. Basically, the weld shall merge smoothly with the parent metal so that the concentration of stress at the weld edge can be minimized.

Weld ripples shall also be reasonably smooth so that if a radiograph from that joint is taken, it can be clearly interpreted. In case the weld surface is irregular or ugly looking, though there are no clear-cut requirements in applicable code sections, at the discretion of the inspection engineer, these welds shall be given a dressing-up grinding. This becomes especially important for inside welds when the tank is expected to carry corrosive fluids. When the weld requires dressings as mentioned, they shall be done prior to radiography of that joint, if this joint is identified for radiography as per requirement.

The random spot radiography specified in code covers mostly pressure contacting welds on the shell alone. However, many client specifications in the oil and gas industry require additional RT, as shown in the previous table. For any radiography taken on a storage tank, the acceptance criteria shall be as provided in API 650 Clause 8.1.5, which refers back to ASME Section VIII UW 51(b) and from there to Appendix 4 of ASME Section VIII Div (1) with regard to rounded indications. These requirements are summarized in Annexure D.

10.3 Fillet Welds (Pressure Retaining)

Three types of fillet welds on the bottom plate fall under this category: lap weld between the annular plate and the bottom plate including three plate lap welds, weld between the shell and the annular plate, and fillet weld of drain-off sump and the like.

Code requires all these welds to be inspected visually, for contour, leg sizes, undercut, and other surface opening defects. In addition to this, other tests (basically the vacuum box test) as specified in code also need to be done to ward off leakages during a hydrostatic test, leading to draining the tank and redoing the hydrostatic test. Fillet weld joints between bottom plates as well as those between the annular plate and the bottom plate and those between the shell and the annular plate are of prime concern, as this weld is expected to take the maximum pressure and overturning load due to acts of nature. Therefore, LPT is specified as a supplementary requirement by many of the client specifications from the oil and gas industry. As far as the fillet welds of gussets that support the roof structure and anchor chairs, the weld shall have the minimum size of fillet mentioned in the drawing. Furthermore, this weld also shall be free from surface defects as mentioned earlier.

Another aspect that is often overlooked by many inspection engineers is the unequal leg size of fillet welds and corner joints in storage tanks. API 650 does not specify any acceptance criteria for the difference in leg size of the fillet welds. As a rule of thumb, a maximum of 2 mm difference between legs is considered acceptable, with the lowest leg having the minimum specified leg size, derived from the fillet size as shown in the drawing. As it is always practical and easy to measure the leg size of a fillet weld rather than the throat size, the measured minimum shall be translated as throat size as required in the applicable drawing.

For roof plate fillet welds, visual and vacuum box test or air leak tests need to be performed. This requirement is applicable to all attachment welds on the roof plate that are expected to make a penetration of the roof plate.

Though it is usual that inspectors allow higher fillet sizes for all fillet welds, in the case of roof plate fillet welds, it shall be controlled strictly to reduce the resultant distortion of the roof plate. Similarly, for frangible roof joints, the size of the fillet has to be controlled precisely as per the drawing.

10.4 Other Structural Welds on Shell and Roof (Inside and Outside)

All structural welds on the entire tank surface shall undergo satisfactory visual inspection. The most important in structural welding is the roof structure welding to the top shell course. Roof structure attachment is decided by service, and it ranges from self-supporting to single or multiple-support column type. The principal consideration for this is the diameter and internal pressures expected inside the tank. Since drawings for the structural steel attachment is decided based on the design loads considered, it shall be the duty of the inspector to ensure the minimum sizes of welds required for each of the components that constitute the roof structure. As live and cyclic loads are also often encountered during service, it is essential that these sizes are maintained strictly according to the drawing or design to the extent possible. The basic minimum requirement is that the attachment proposed shall be capable of supporting the roof and also should be sound enough to withstand the upward forces that start acting on the roof while the tank is in service.

Since the design of the weld joints of roof structure to shell and those welds between structural members is based on the above considerations, compliance to required fillet size with reasonable contour needs to be ensured during construction. For fillet welds of gussets that support the roof structure, weld shall have the minimum size of fillet mentioned in the drawing. Furthermore, this weld also shall be free from undercuts and other surface defects as mentioned earlier.

10.5 Inspection (Dimensional)

10.5.1 Annular and Bottom plates

Dimensions

The annular plates shall be laid generally as per the dimensions in the drawing. The most critical dimension specified in the code against this is the minimum width of the annular plate. As per API 650, the minimum width requirement for this plate is 600 mm from the inside of the shell and the lap weld between the annular plate and the bottom plate (see Clause 5.1.5.4. of API 650). However, higher widths for annular plates are required by many client specifications or in the data sheets enclosed with inquiry. Therefore, the only parameter to be measured on the annular plate is the minimum width, apart from the thickness of the plate used.

Similarly, the dimensions of the bottom plates also shall be checked according to that specified in the drawing. Generally, the size of the annular and bottom plates shall be based on the availability of the specific plate material proposed for tank construction, as well as logistic constraints. Because of this, no specific tolerance is specified in this regard. However, for reducing the quantum of weld length, it is important to go for the highest plate size possible.

Buckling of the Bottom Plate

Yet another aspect to be inspected and recorded is the buckling of bottom plates after welding. Buckling of the bottom plate is an unavoidable defect arising out of shrinkages taking place in the huge quantity of welding involved in bottom plate fabrication. Furthermore, the low thickness of bottom plates usually provided also adds to this distortion. By adopting a proper welding sequence, this can be controlled to a great extent (see the welding sequence recommended in Chapter 7, Section 7.8.9). Clause 7.3.6.8 of API 650 specifies the methodology to record distortion or buckling of the bottom plate. This measurement is required to be carried out before and after the hydrostatic test of the tank. This would be of immense help in cross-verification with settlement (planar and out of plane) measurements taken during planned shutdowns during the service life of the tank to arrive at a rational decision with reliability about the extent of repairs required during shutdowns for the extended service life.

Bottom plate buckling shall be recorded after completion of all bottom plate welding, including full welding of the shell to the annular plate and that of the clean-out door and its components. Tank bottom buckling shall be recorded using an even number of diameter lines established at 10 m (32 ft) intervals over the circumference, with a minimum of four such diameters.

Similarly, along diameter lines, measurements shall be taken at 3 m (10 ft) intervals or eight measurements as a minimum, as shown in the following sketch.

Buckling measurement inside the tank shall be recorded with reference to a reference point transferred to the inside surface of the shell (say at 200 mm from the surface of the bottom plate) and marked on the shell using permanent markers, which can be used for taking buckling/settlement measurements after hydrostatic testing and during periodic turnaround as well. This reference point shall be transferred to all locations identified over the circumference using water level. Buckling across a diameter is recorded by measuring the depth to the surface of the bottom plate at 3 m (10 ft) intervals as shown in Figure 10.1 along the diameters identified.

Though the methodology to record buckling of the bottom plate is specified in API 650, it does not specify any tolerance for the same. In the event of a concrete ring wall foundation, used commonly in the oil and gas industry, the level variation permitted for the foundation is ±3 mm (1/8″) in any 3 m (10′) of the circumference and within ±6 mm (1/4″) in total circumference measured from average elevation. So tanks constructed on a concrete ring wall foundation shall have a buckling tolerance above this level to accommodate distortion of the bottom plate due to welding. Based on experience from the field, a rational level of buckling tolerance achievable for tanks with a concrete ring wall foundation is ±12 mm (1/2″) in total circumference.

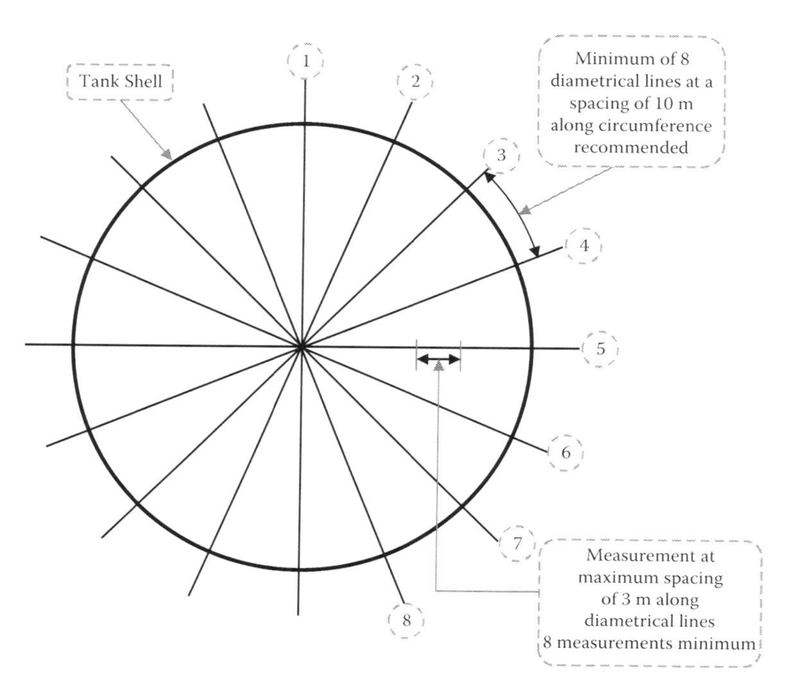

FIGURE 10.1
Measurement of tank bottom plate buckling.

10.5.2 Shell

Apart from basic dimensions such as outside circumference, outside diameter, and so on, API 650 specifies a minimum width of 1,800 mm for shell plates (Clause 5.6.1.2 of API 650). Moreover, shell plates need to be curved to the required diameter by any suitable method as indicated in Section 7.7.2 in Chapter 7. The most efficient and effective method is bending the plates in plate bending machines. As mentioned earlier, forming the plate to the required diameter is not a must for all thicknesses. The diameter thickness combinations requiring plate bending are given in Clause 6.1.3 of API 650.

Misalignment of Shells in Vertical Joints

Shell plates that are to be joined shall be matched precisely and retained in position during the welding operation. Though the target is to attain perfect matching of surfaces, in practice it is not possible, and hence tolerances are required for every dimension. The tolerance on misalignment is usually related to the thickness of the shell, and as a rule of thumb it is taken as 10% of the shell with an upper limit of 2 mm. API Clause 7.2.3.1 specifies this requirement as 1.5 mm (1/16″) up to shell thickness up to 16 mm and thereafter at 10% of shell thickness with a maximum of 3 mm (1/8″).

Misalignment of Shells in Horizontal Joints

For horizontal butt welds, the tolerance is specified in API 650 (Clause 7.2.3.2). However, it is stated in a slightly different manner as the projection of the top plate beyond the face of the lower plate rather than just a mismatch over a calculated difference, which occurs when shell thicknesses vary at horizontal joints. Furthermore, the standard calls for tapering the edges when mismatch is more than 3 mm.

Roundness

Yet another measurement to be verified at the site is the roundness of the shells. Clause 7.5.3 of API 650 requires the radii to be measured at 300 mm (1′) above the corner weld, and tolerances depending on tank diameters is provided in the table provided under this clause, which range from ±13 mm (1/2″) to ±32 mm (1¼″).

Circumference at Each Shell Course

API 650 is silent about this measurement. However, the following considerations make this measurement so important and critical.

When each shell course is set up, the outside circumference needs to be taken, from which the mean outside or inside diameter can be calculated. The roundness measure needs to be verified with the diameter calculated in this manner as well and shall be within limits for the bottom shell course.

For the subsequent shells, the circumference needs to be recorded at both edges of the shell courses (say at 100 mm to 150 mm from the weld edge). For the misalignment tolerance to be within acceptance limits as specified in API 650 Clause 7.2.3.2, the circumference shall be within ±9.5 mm, for mating shells of the same thickness, with a permitted misalignment of 1.5 mm based on plate thickness. Table 10.2 with a sample calculation under a different scenario explains this requirement in detail.

It is almost impossible to distribute the mismatch present in every horizontal joint uniformly around the circumference on one side, either inside or outside. In all practical cases, it will vary from –1.5 to +1.5 over the entire

TABLE 10.2

Misalignment Calculation

Location ID	Tank Inside Diameter (mm)	Shell Thickness (mm)	Required Outside Circumference (mm)
Bottom shell top	24,000	12	75,435.3
Second shell bottom	24,000	12 + 1.5 mismatch all around	75,444.8

Note: Maximum possible variation in circumference is only ±9.5 mm over circumference of the previous shell course.

circumference, resulting in less variation in circumference. When the set up of the shell plates is carried out progressively with the mismatch as specified, the circumference measurements shall be within the calculated values as shown previously.

When shell course thicknesses vary, from bottom to top, the calculation shall take care of the thickness difference as well, along with permitted mismatch based on the thickness of the shell.

Plumbness

API 650 provides tolerance on plumbness for erected and welded shell courses from top to bottom and for individual shells. It also provides tolerance for plumbness for roof columns, guide poles, and other vertical components, especially those that traverse the entire height of the tank. The tolerance permitted is 1 in 200 of the total tank height, both for erected shell courses and for roof support columns, guide poles, and so on for fixed cone roof tanks (Clause 7.5.2 of API 650).

For plumbness in individual shell courses, the tolerance is the same as permissible variations for flatness and waviness specified for bare plates used for construction. For the ASTM materials specified in API 650, this would necessitate reference to ASTM A 6, ASTM A 20, or ASTM A 480, depending on the respective ASTM specification to which plates are procured.

Though the above specified is the requirement as per API 650 for local deviations, some client specifications set explicit limits for the same as follows based on the thickness of the shell sections.

In the event of floating roof tanks, as per API 650, the tolerance on plumbness for shells remains the same. Appurtenances such as columns, ladders, and antirotation devices that penetrate the deck need to be within 1 in 200 of total height, limited to ±75 mm (±3″).

Local Deviations

Clause 7.5.4 of API 650 specifies tolerance for local deviations from theoretical shape, including flat spots, peaking and banding of welds, etc. Peaking at vertical joints and banding at horizontal joints are limited to ±13 mm (1/2″).

TABLE 10.3

Tolerance on Local Deviations (Restriction by Clients)

	Tolerance	
Plate Thickness	**Fixed Roof Tanks**	**Floating Roof Tanks**
Plates ≤ 12.5 mm thickness	16 mm	10 mm
Plates > 12.5 mm to ≤ 25 mm thickness	13 mm	8 mm
Plates > 25 mm thickness	10 mm	6 mm

For the measurement of peaking at vertical weld joints and banding at horizontal weld joints, horizontal (made to required curvature) and vertical sweep boards (straight edge type) of 900 mm (36″) length is proposed by API 650.

Moreover, for flat spots measured in vertical plane, the limits specified in API 650 refers to flatness or waviness specified in ASTM A 6, ASTM A 20, or ASTM A 480, as referred in respective ASTM specification for shell plates used in construction.

For columns, ladders, and other rigid vertical appurtenances that penetrate the deck and provided with a seal, in floating roof tanks, the tolerance permitted for local deviation is ±125 mm (±5″). (Clause H 4.5 of API 650).

For all local departures from the design form for shell, some client specifications usually restrict the tolerance limits further based on the thickness of shell courses. Local departure in form, measured horizontally and vertically shall not exceed the following, when measured over a gauge length of 2500 mm (100″), reasonably away from weld seams:

Such departures from the designed form shall be gradual over the gauge length, sharp changes in form are not permitted.

Nozzles and Manholes

Clauses 7.5.6 and 7.5.7 of API 650 specify clear-cut tolerances for completed nozzles and manways as follows.

Many of the client specifications (even those from the oil and gas industry) are in agreement with the tolerances provided in API 650, and hence no more restrictions are specifically provided.

TABLE 10.4

Tolerance on Nozzle Dimensions

Description of Dimension	Tolerance
Projection from the outside of the tank shell to the extreme face of the flange (raised face or flat face)	±5 mm (3/16″)
Elevation of the shell nozzle or radial location of a roof nozzle (though measurement is up to the center of the nozzle pipe but measured at the nearest outside diameter of the pipe for convenience)	±6 mm (1/4″)
Flange tilt in any plane, measured on the flange face	±1/2 degree for nozzles greater than NPS 300 (12″) ±3 mm (1/8″) at the outside flange diameter for nozzles NPS 300 (12″) and smaller
Flange bolt hole orientation (shall straddle centerline as a general practice unless otherwise specified)	±3 mm (1/8″)

Similarly, for manways coming on the shell, the tolerance on projection, angular position, elevation, and flange tilt is required to be within ±13 mm (1/2″). However, it is observed that it is easy to maintain this measurement within ±10 mm (3/8″) in practice.

API 650 is silent on tolerance for projection, angular position, through and angular tilt of roof manways. However, the tolerance specified for shell manways is achievable for various dimensions of manways coming on the roof plate. In the event of roof nozzles and manways, because of the possibility of buckling the roof (due to welding and lower thickness), taking realistic measurements is a bit tricky.

11

Nondestructive Testing

11.1 Radiographic Testing

The most extensively used nondestructive testing (NDT) method in tank construction is radiography. Conventionally, gamma radiography is performed on storage tanks. In gamma radiography, Iridium 192 source is the most commonly used source in the industry. In some specific instances such as the use of some special highly productive welding processes such as electro gas welding, some clients ask for X-ray radiography. For both of these radiography tests, the applicable code for carrying out radiography is ASME Section V. The extent of radiography and its acceptance criteria are provided in Section 8 of API650, the excerpts of which are given below.

11.2 Minimum Number and Location of Radiographs Required as per API 650

Consideration

1 For arriving at number, location, and type of radiograph required, adjoining shell plates are considered to be of the same thickness, up to a thickness difference of 3 mm (1/8″).

Welds Requiring No Radiography

2 Roof plate welds, bottom plate welds, welds joining the top angle to either the roof or the shell, welds joining the shell plate to the bottom plate, welds in the nozzle and manway necks made from the plate, or appurtenance welds to the tank.

Welds Requiring Radiography

3 Shell butt welds (as referred in Clauses 8.1.2.2, 8.1.2.3, and 8.1.2.4), annular plate butt welds (Clause 8.1.2.9), and flush-type connections with butt welds (Clause 5.7.8.11).

For Butt Welded Vertical Joints Wherein the Thinner Shell Plate Is ≤10 mm (3/8″) Thick

4 One spot radiograph in the first 3 m (10′) of the completed vertical joint of each type and thickness welded by each welder or welding operator, followed by one spot per each additional 30 m (100′) vertical weld or part thereof, regardless of the number of welders used. Of this, at least 25% of spots shall be at junctions of vertical and horizontal joints, with a minimum of two such intersections per tank. In addition, one random spot radiograph shall be taken in each vertical joint in the lowest course.

For Butt Welded Vertical Joints Wherein the Thinner Shell Plate Is >10 mm (3/8″) but ≤25 mm (1″) Thick

5 Spots as specified in Item 4 and all junctions of vertical and horizontal joints in plates of this thickness range.[a]

6 In the lowest course, two spot radiographs shall be taken in each vertical joint, with one as close to the bottom as is practicable, and the other shall be taken at random.

Vertical Joints in Shell Plates of Thickness >25 mm (1″)

7 One hundred percent radiography; all junctions of vertical and horizontal joints in this thickness range are included.

Butt Weld around the Periphery of Insert Plates

8 Weld that extends less than the adjacent shell course height and that contains shell openings (i.e., nozzle, manway, flush-type clean out, flush-type shell connection) and their reinforcing elements require 100% radiography.

9 Weld that extends to match the adjacent shell course height shall have the vertical and the horizontal butt joints and the intersections of vertical and horizontal weld joints radiographed using the same rules that apply to the weld joints in adjacent shell plates in the same shell course.

Horizontal Joints in Shell Plates

10 One spot radiograph in the first 3 m (10′) of the completed horizontal butt joint of the same type and thickness (based on the thickness of the thinner plate at the joint) regardless of the number of welders or welding operators, followed by one radiograph in each additional 60 m (200′) (approximately) and part thereof. These radiographs are in addition to the radiographs of junctions of vertical joints indicated in Items 4, 5, and 6.

Bottom Annular Plate Butt Welds (If Clauses 5.5.1 or M 4.1 Are Applicable)[b, c]

11 For double-welded butt joints, one spot radiograph is required on 10% of the radial joints.

12 For single-welded butt joints with a permanent or removable backup bar, one spot radiograph is required on 50% of the radial joints.

Notes:

[a] Radiographs at "T" joints shall clearly show not less than 75 mm (3″) of vertical weld and 50 mm (2″) of weld length on each side of the vertical intersection (to be read in conjunction with Item 5).

[b] Each radiograph shall clearly show a minimum of 150 mm (6″) of weld length. The film shall be centered on the weld and shall be of sufficient width to permit adequate space for the location of identification marks and an image quality indicator (IQI).

[c] Locations of radiographs shall preferably be at the outer edge of the joint where the shell plate and annular plate join.

[d] The requirement of the number of spot radiographs is based on a per tank basis, irrespective of the number of tanks being erected concurrently or continuously at any location.

e It is recognized that in many cases the same welder or welding operator does not weld both sides of a butt joint. If two welders or welding operators weld opposite sides of a butt joint, it is permissible to examine their work with one spot radiograph. If the radiograph is rejected, additional spot radiographs shall be taken to determine whether one or both of the welders or welding operators are at fault.

f An equal number of spot radiographs shall be taken from the work of each welder or welding operator in proportion to the length of joints welded.

Based on these requirements, the shell radiography requirement using varying thicknesses for the shell for a typical storage tank is provided in the following table for easy understanding, covering almost all the conditions specified in API 650. Please note that the requirements are purely based on API 650 rounded off to the next positive integer with regard to numbers. While arriving at this pictorial presentation, the following assumptions were made:

1	Plate size used is 12,000 mm × 2,500 mm for all thickness ranges indicated.
2	Five welders are assigned to carry out the complete welding of all vertical and horizontal seams.
3	Extra welders are to be added for carrying out other structural welding of the tank.
4	In case of tight schedules, more welders may have to be employed, which may have a slight impact on the total radiography requirements mentioned in the sketches overleaf.
5	The same welder is expected to complete welding from both sides of a particular seam to the extent possible.
6	The same welders engaged for vertical seams shall be employed for welding the horizontal seams also.
7	Welders shall be given a predesignated length of weld in any horizontal joint based on the staggering required while welding. The same welders shall cover welding from both inside and outside in predesignated lengths of welds assigned to them.

11.3 Weld Maps with Pictorial Presentation of Radiography

Weld Map with Spot Radiography (Shell thickness 22-8 mm)

Diameter	Height	Plate size	Type of joining of shells
38,000 mm	20,000 mm	12,000mm X 2500 mm	Inside flush

Shell thickness	Shell 1	Shell 2	Shell 3	Shell 4	Shell 5	Shell 6	Shell 7	Shell 8
	22 mm	20 mm	18 mm	12 mm	10 mm	8 mm	8 mm	8 mm

FIGURE 11.1A
Refer to Table 11.1.

Weld Map with Spot Radiography (Shell Thickness 28-8mm)

Diameter	Height	Plate size	Type of joiningof shells
38,000 mm	20,000 mm	12,000mm X 2500 mm	Inside flush

Shell thickness	Shell 1	Shell 2	Shell 3	Shell 4	Shell 5	Shell 6	Shell 7	Shell 8
	28 mm	25 mm	22 mm	19 mm	16 mm	13 mm	10 mm	8 mm

FIGURE 11.1B
Refer to Table 11.2.

11.3.1 Table 11.1 for Pictorial Presentation of Radiography (Shell Thickness 22 mm to 8 mm)

TABLE 11.1

Radiography Spot Selection (Example for Shell Thickness 22 mm to 8 mm)

Summary of Radiography Requirement Based on Weld Length and Welders Deployed								
Serial No.	**Weld Seam**	**Type of Weld**	**Length (m)**	**Welder No.**	**Thickness (mm)**	**No. of RT Spots Marked**	**Remarks**	**Explanation Notes**
1	S1 V1	Butt	2.5	W1	22	3	Closing joint	a, c, d
2	S1 V2	Butt	2.5	W1	22	3		b, c, d
3	S1 V3	Butt	2.5	W2	22	3		a, c, d
4	S1 V4	Butt	2.5	W2	22	2		c, d
5	S1 V5	Butt	2.5	W3	22	3		a, c, d
6	S1 V6	Butt	2.5	W3	22	2		c, d
7	S1 V7	Butt	2.5	W4	22	3		a, c, d
8	S1 V8	Butt	2.5	W4	22	2		c, d
9	S1 V9	Butt	2.5	W5	22	3		a, c, d
10	S1 V10	Butt	2.5	W5	22	2		c, d
11	S2 V1	Butt	2.5	W2	20	3	Closing joint	a, d
12	S2 V2	Butt	2.5	W2	20	3		d, e
13	S2 V3	Butt	2.5	W3	20	3		a, d
14	S2 V4	Butt	2.5	W3	20	2		d
15	S2 V5	Butt	2.5	W4	20	3		a, d
16	S2 V6	Butt	2.5	W4	20	2		d
17	S2 V7	Butt	2.5	W5	20	3		a, d
18	S2 V8	Butt	2.5	W5	20	2		d
19	S2 V9	Butt	2.5	W1	20	3		a, d
20	S2 V10	Butt	2.5	W1	20	2		d
21	S3 V1	Butt	2.5	W3	18	3	Closing joint	a, d
22	S3 V2	Butt	2.5	W3	18	3		d, f
23	S3 V3	Butt	2.5	W4	18	3		a, d
24	S3 V4	Butt	2.5	W4	18	2		d
25	S3 V5	Butt	2.5	W5	18	3		a, d
26	S3 V6	Butt	2.5	W5	18	2		d
27	S3 V7	Butt	2.5	W1	18	3		a, d
28	S3 V8	Butt	2.5	W1	18	2		d
29	S3 V9	Butt	2.5	W2	18	3		a, d
30	S3 V10	Butt	2.5	W2	18	2		d
31	S4 V1	Butt	2.5	W4	12	3	Closing joint	a, d
32	S4 V2	Butt	2.5	W4	12	3		d
33	S4 V3	Butt	2.5	W5	12	3		a, d
34	S4 V4	Butt	2.5	W5	12	2		d, g

(Continued)

TABLE 11.1

Radiography Spot Selection (Example for Shell Thickness 22 mm to 8 mm) (*Continued*)

Summary of Radiography Requirement Based on Weld Length and Welders Deployed

Serial No.	Weld Seam	Type of Weld	Length (m)	Welder No.	Thickness (mm)	No. of RT Spots Marked	Remarks	Explanation Notes
35	S4 V5	Butt	2.5	W1	12	3		a, d
36	S4 V6	Butt	2.5	W1	12	2		d
37	S4 V7	Butt	2.5	W2	12	3		a, d
38	S4 V8	Butt	2.5	W2	12	2		d
39	S4 V9	Butt	2.5	W3	12	3		a, d
40	S4 V10	Butt	2.5	W3	12	2		d
41	S5 V1	Butt	2.5	W5	10	3	Closing joint	a, d
42	S5 V2	Butt	2.5	W5	10	3		d, h
43	S5 V3	Butt	2.5	W1	10	3		a, d
44	S5 V4	Butt	2.5	W1	10	2		d
45	S5 V5	Butt	2.5	W2	10	3		a, d
46	S5 V6	Butt	2.5	W2	10	2		d
47	S5 V7	Butt	2.5	W3	10	3		a, d
48	S5 V8	Butt	2.5	W3	10	2		d
49	S5 V9	Butt	2.5	W4	10	3		a, d
50	S5 V10	Butt	2.5	W4	10	2		d
51	S6 V1	Butt	2.5	W1	8	1	Closing joint	i
52	S6 V2	Butt	2.5	W1	8			
53	S6 V3	Butt	2.5	W2	8	1		i
54	S6 V4	Butt	2.5	W2	8			
55	S6 V5	Butt	2.5	W3	8	1		i
56	S6 V6	Butt	2.5	W3	8			
57	S6 V7	Butt	2.5	W4	8	1		i
58	S6 V8	Butt	2.5	W4	8			
59	S6 V9	Butt	2.5	W5	8	1		i
60	S6 V10	Butt	2.5	W5	8			
61	S7 V1	Butt	2.5	W2	8		Closing joint	
62	S7 V2	Butt	2.5	W2	8			
63	S7 V3	Butt	2.5	W3	8	1		j
64	S7 V4	Butt	2.5	W3	8			
65	S7 V5	Butt	2.5	W4	8			
66	S7 V6	Butt	2.5	W4	8			
67	S7 V7	Butt	2.5	W5	8			
68	S7 V8	Butt	2.5	W5	8	1		j
69	S7 V9	Butt	2.5	W1	8			
70	S7 V10	Butt	2.5	W1	8			

TABLE 11.1

Radiography Spot Selection (Example for Shell Thickness 22 mm to 8 mm) (*Continued*)

Summary of Radiography Requirement Based on Weld Length and Welders Deployed

Serial No.	Weld Seam	Type of Weld	Length (m)	Welder No.	Thickness (mm)	No. of RT Spots Marked	Remarks	Explanation Notes
71	S8 V1	Butt	2.5	W3	8		Closing joint	
72	S8 V2	Butt	2.5	W3	8			
73	S8 V3	Butt	2.5	W4	8			
74	S8 V4	Butt	2.5	W4	8			
75	S8 V5	Butt	2.5	W5	8	1		k
76	S8 V6	Butt	2.5	W5	8			
77	S8 V7	Butt	2.5	W1	8			
78	S8 V8	Butt	2.5	W1	8			
79	S8 V9	Butt	2.5	W2	8			
80	S8 V10	Butt	2.5	W2	8		Closing joint	
81	H1	Fillet	120					l
82	H2	Butt	120		22/20	3		m
83	H3	Butt	120		20/18	3		n
84	H4	Butt	120		18/12	3		o
85	H5	Butt	120		12/10	3		p
86	H6	Butt	120		10/8	3		q
87	H7	Butt	120		8/8	3		r
88	H8	Butt	120		8/8	3		s
89	H9	Butt	120		8/8	3		t

Note: 1. Five welders are engaged in the welding of the shell as indicated in the summary.
2. Welding from inside and outside of the seam will be carried out by the same welder.
3. The same welders will be welding the horizontal seams also, as it is done after the vertical seam welding.
4. The same welders are required for the construction of the storage tank, and they shall be kept occupied by the manufacturer on other works.
5. A plate size of 12 m length and 2.5 m width was assumed, with 10 shell courses of the same height.
6. The circumference works out to 120 m, and for horizontal seams, each welder shall weld 24 m so that all 5 can complete 120 m.
7. While welder W1 starts at S1 V1 for H2, welder W2 starts welding H2 at S2 V1, and so on. Repeating cycle at H7.
8. Joint H9 (between S8 and curb angle) is considered a butt joint.

[a] **Green** middle spot: Based on Clause 8.1.2.2(b) implemented through 8.1.2.2(a). First 3 m of each welder, provided as random as T joints are covered by Clause 8.1.2.2(b).

[b] **Green** middle spot: Based on Clause 8.1.2.2(b) implemented through Clause 8.1.2.2(a). Additional spot for remaining 12.5 m of vertical weld S1 V2, V4, V6, V8, and V10 irrespective of welder, which coincides with Note c.

[c] **Red** bottom spot: Based on Clause 8.1.2.2(b). Two random spots, one close to the bottom plate weld.

(*Continued*)

TABLE 11.1

Radiography Spot Selection (Example for Shell Thickness 22 mm to 8 mm) (*Continued*)

[d] **Red** top and bottom T joint spots: Clause 8.1.2.2(b). All junctions between vertical joints and horizontal joints.
[e] **Green m**iddle spot: Clause 8.1.2.2(a). Additional spot for remaining 12.5 m of vertical welds S2 V2, V4, V6, V8, and V10 irrespective of welder marked for S2 V2.
[f] **Green m**iddle spot: Clause 8.1.2.2(a). Additional spot for remaining 12.5 m of vertical welds S3 V2, V4, V6, V8, and V10 irrespective of welder marked for S3 V2.
[g] **Green** middle spot: Clause 8.1.2.2(a). Additional spot for remaining 12.5 m of vertical welds S4 V2, V4, V6, V8, and V10 irrespective of welder marked for S4 V2.
[h] **Green m**iddle spot: Clause 8.1.2.2(a). Additional spot for remaining 12.5 m of vertical welds S5 V2, V4, V6, V8, and V10 irrespective of welder marked for S5 V2.
[i] **Yellow** spots: Clause 8.1.2.2(a). First 3 m of each welder, provided as random with two at T joints to comply to requirement of 25% on T joints S6 V3 and S6 V9, W1 and W4, respectively.
[j] **Yellow** spots: Clause 8.1.2.2(a). Additional spot (3 nos) for remaining 62.5 m of vertical weld in shell courses S6, S7, and S8 required. Two provided in shell course S7 with one at T joint to meet 25% requirement in T. Remaining one provided in shell S8.
[k] Remaining one **yellow** spot of Note j provided at S8 V5 random location, as the horizontal joint between curb angle and shell is not considered critical on account of low liquid level possible.
[l] Bottom plate to shell weld. No radiography is required.
[m] As per Clause 8.1.2.3, randomly distributed to welder zones of W1, W2, and W4 in H2 (S1/S2).
[n] As per Clause 8.1.2.3, randomly distributed to welder zones of W2, W3, and W5 in H3 (S2/S3).
[o] As per Clause 8.1.2.3, randomly distributed to welder zones of W3, W4, and W1 in H4 (S3/S4).
[p] As per Clause 8.1.2.3, randomly distributed to welder zones of W4, W5, and W2 in H5 (S4/S5).
[q] As per Clause 8.1.2.3, randomly distributed to welder zones of W5, W1, and W3 in H6 (S5/S6).
[r] As per Clause 8.1.2.3, randomly distributed to welder zones of W1, W2, and W4 in H7 (S6/S7).
[s] As per Clause 8.1.2.3, randomly distributed to welder zones of W2, W3, and W5 in H8 (S7/S8).
[t] As per Clause 8.1.2.3, randomly distributed to welder zones of W3, W4, and W1 in H9 (S8/Curb angle).

11.3.2 Table 11.2 for Pictorial Presentation of Radiography (Shell Thickness 28 mm to 8 mm)

TABLE 11.2

Radiography Spot Selection (Example for Shell Thickness 28 mm to 8 mm)

Summary of Radiography Requirement Based on Weld Length and Welders Deployed								
Serial No.	Weld Seam	Type of Weld	Length (m)	Welder No.	Thickness (mm)	No. of RT Spots Marked	Remarks	Explanation Notes
1	S1 V1	Butt	2.5	W1	28	100%	Closing joint	a
2	S1 V2	Butt	2.5	W1	28	100%		
3	S1 V3	Butt	2.5	W2	28	100%		
4	S1 V4	Butt	2.5	W2	28	100%		
5	S1 V5	Butt	2.5	W3	28	100%		

TABLE 11.2

Radiography Spot Selection (Example for Shell Thickness 28 mm to 8 mm)

Summary of Radiography Requirement Based on Weld Length and Welders Deployed								
Serial No.	**Weld Seam**	**Type of Weld**	**Length (m)**	**Welder No.**	**Thickness (mm)**	**No. of RT Spots Marked**	**Remarks**	**Explanation Notes**
6	S1 V6	Butt	2.5	W3	28	100%		
7	S1 V7	Butt	2.5	W4	28	100%		
8	S1 V8	Butt	2.5	W4	28	100%		
9	S1 V9	Butt	2.5	W5	28	100%		
10	S1 V10	Butt	2.5	W5	28	100%		
11	S2 V1	Butt	2.5	W2	25	100%	Closing joint	b
12	S2 V2	Butt	2.5	W2	25	100%		
13	S2 V3	Butt	2.5	W3	25	100%		
14	S2 V4	Butt	2.5	W3	25	100%		
15	S2 V5	Butt	2.5	W4	25	100%		
16	S2 V6	Butt	2.5	W4	25	100%		
17	S2 V7	Butt	2.5	W5	25	100%		
18	S2 V8	Butt	2.5	W5	25	100%		
19	S2 V9	Butt	2.5	W1	25	100%		
20	S2 V10	Butt	2.5	W1	25	100%		
21	S3 V1	Butt	2.5	W3	22	3	Closing joint	c, d
22	S3 V2	Butt	2.5	W3	22	3		c, d
23	S3 V3	Butt	2.5	W4	22	3		c, e
24	S3 V4	Butt	2.5	W4	22	2		d
25	S3 V5	Butt	2.5	W5	22	3		c, d
26	S3 V6	Butt	2.5	W5	22	2		d
27	S3 V7	Butt	2.5	W1	22	3		c, d
28	S3 V8	Butt	2.5	W1	22	2		d
29	S3 V9	Butt	2.5	W2	22	3		c, d
30	S3 V10	Butt	2.5	W2	22	2		d
31	S4 V1	Butt	2.5	W4	19	3	Closing joint	c, d
32	S4 V2	Butt	2.5	W4	19	3		c, e
33	S4 V3	Butt	2.5	W5	19	3		c, d
34	S4 V4	Butt	2.5	W5	19	2		d
35	S4 V5	Butt	2.5	W1	19	3		c, d
36	S4 V6	Butt	2.5	W1	19	2		d
37	S4 V7	Butt	2.5	W2	19	3		c, d
38	S4 V8	Butt	2.5	W2	19	2		d
39	S4 V9	Butt	2.5	W3	19	3		c, d
40	S4 V10	Butt	2.5	W3	19	2		d

(Continued)

TABLE 11.2

Radiography Spot Selection (Example for Shell Thickness 28 mm to 8 mm) (*Continued*)

Summary of Radiography Requirement Based on Weld Length and Welders Deployed

Serial No.	Weld Seam	Type of Weld	Length (m)	Welder No.	Thickness (mm)	No. of RT Spots Marked	Remarks	Explanation Notes
41	S5 V1	Butt	2.5	W5	16	3	Closing joint	c, d
42	S5 V2	Butt	2.5	W5	16	3		c, e
43	S5 V3	Butt	2.5	W1	16	3		c, d
44	S5 V4	Butt	2.5	W1	16	2		d
45	S5 V5	Butt	2.5	W2	16	3		c, d
46	S5 V6	Butt	2.5	W2	16	2		e
47	S5 V7	Butt	2.5	W3	16	3		c, d
48	S5 V8	Butt	2.5	W3	16	2		d
49	S5 V9	Butt	2.5	W4	16	3		c, d
50	S5 V10	Butt	2.5	W4	16	2		d
51	S6 V1	Butt	2.5	W1	13	3	Closing joint	c, d
52	S6 V2	Butt	2.5	W1	13	3		c, e
53	S6 V3	Butt	2.5	W2	13	3		c, d
54	S6 V4	Butt	2.5	W2	13	2		d
55	S6 V5	Butt	2.5	W3	13	3		c, d
56	S6 V6	Butt	2.5	W3	13	2		d
57	S6 V7	Butt	2.5	W4	13	3		c, d
58	S6 V8	Butt	2.5	W4	13	2		d
59	S6 V9	Butt	2.5	W5	13	3		c, d
60	S6 V10	Butt	2.5	W5	13	2		d
61	S7 V1	Butt	2.5	W2	10	3	Closing joint	c, d
62	S7 V2	Butt	2.5	W2	10	3		c, e
63	S7 V3	Butt	2.5	W3	10	3		c, d
64	S7 V4	Butt	2.5	W3	10	2		d
65	S7 V5	Butt	2.5	W4	10	3		c, d
66	S7 V6	Butt	2.5	W4	10	2		d
67	S7 V7	Butt	2.5	W5	10	3		c, d
68	S7 V8	Butt	2.5	W5	10	2		d
69	S7 V9	Butt	2.5	W1	10	3		c, d
70	S7 V10	Butt	2.5	W1	10	2		d
71	S8 V1	Butt	2.5	W3	8	1	Closing joint	f
72	S8 V2	Butt	2.5	W3	8	1		f, g
73	S8 V3	Butt	2.5	W4	8	1		f, h
74	S8 V4	Butt	2.5	W4	8			
75	S8 V5	Butt	2.5	W5	8	1		f

TABLE 11.2

Radiography Spot Selection (Example for Shell Thickness 28 mm to 8 mm) (*Continued*)

Summary of Radiography Requirement Based on Weld Length and Welders Deployed

Serial No.	Weld Seam	Type of Weld	Length (m)	Welder No.	Thickness (mm)	No. of RT Spots Marked	Remarks	Explanation Notes
76	S8 V6	Butt	2.5	W5	8			
77	S8 V7	Butt	2.5	W1	8	1		f, h
78	S8 V8	Butt	2.5	W1	8			
79	S8 V9	Butt	2.5	W2	8	1		f
80	S8 V10	Butt	2.5	W2	8			
81	H1	Fillet	120					i
82	H2	Butt	120		25/22	3		j
83	H3	Butt	120		25/22	3		k
84	H4	Butt	120		22/19	3		l
85	H5	Butt	120		19/16	3		m
86	H6	Butt	120		16/13	3		n
87	H7	Butt	120		13/10	3		o
88	H8	Butt	120		10/8	3		p
89	H9	Butt	120		8/8	3		q

Note: 1. Five welders are engaged in the welding of the shell as indicated in the summary.
2. Welding from inside and outside of the seam will be carried out by the same welder.
3. The same welders will be welding the horizontal seams also, as it is done after the vertical seam welding.
4. The same welders are required for the construction of the storage tank, and they shall be kept occupied by the manufacturer on other works.
5. A plate size of 12 m length and 2.5 m width was assumed, with 10 shell courses of the same height.
6. The circumference works out to 120 m, and for horizontal seams, each welder shall weld 24 m so that all 5 can complete 120 m.
7. While welder W1 starts at S1 V1 for H2, welder W2 starts welding H2 at S2 V1, and so on. Repeating cycle at H7.
8. Joint H9 (between S8 and curb angle) is considered as a butt joint.

[a] One hundred percent RT as per Clause 8.1.2.2(c) including T joints.
[b] One hundred percent RT as per Clause 8.1.2.2(c) including T joints at both ends.
[c] Middle spot based on Clause 8.1.2.2(b) implemented through Clause 8.1.2.2(a).
[d] Additional spots at both T joints as per Clause 8.1.2.2(b).
[e] Additional spot as per Clause 8.1.2.2(a) for remaining 12.5 m of vertical weld.
[f] Spot as per Clause 8.1.2.2(a). One spot per every welder for the first 3 m.
[g] Additional spot for remaining 12.5 m of weld as per Clause 8.1.2.2(a).
[h] Twenty-five percent of selected spots shall be at T joints as per Clause 8.1.2.2(a).
[i] Bottom plate to shell weld. No radiography is required.
[j] As per Clause 8.1.2.3, randomly distributed to welder zones of W1, W2, and W4 in H2 (S1/S2).
[k] As per Clause 8.1.2.3, randomly distributed to welder zones of W2, W3, and W5 in H3 (S2/S3).
[l] As per Clause 8.1.2.3, randomly distributed to welder zones of W3, W4, and W1 in H4 (S3/S4).
[m] As per Clause 8.1.2.3, randomly distributed to welder zones of W4, W5, and W2 in H5 (S4/S5).
[n] As per Clause 8.1.2.3, randomly distributed to welder zones of W5, W1, and W3 in H6 (S5/S6).
[o] As per Clause 8.1.2.3, randomly distributed to welder zones of W1, W2, and W4 in H7 (S6/S7).
[p] As per Clause 8.1.2.3, randomly distributed to welder zones of W2, W3, and W5 in H8 (S7/S8).
[q] As per Clause 8.1.2.3, randomly distributed to welder zones of W3, W4, and W1 in H9 (S8/Curb angle).

11.4 Other Requirements for Radiography

11.4.1 Technique

Many times, client specifications add restrictions to the technique and type of films to be used for radiography. In addition, based on the service requirements, the extent of radiography required is also increased beyond those specified in API 650. Some of the salient additions made from the oil and gas industry related to the storage of hydrocarbons are indicated in the summary of inspection and tests provided in Chapter 10 (Section 10.1). In the absence of any such specification, basic minimum requirements in API 650 shall be complied with regarding the extent of testing and ASME Section V, Article 2 regarding the technique to be employed. (Refer to Clause 8.1.3.1 of API 650.)

11.4.2 Personnel

Personnel performing radiography and evaluating radiographs shall be qualified by the American Society for Nondestructive Testing (ASNT) SNT-TC-1A Level II or III.

11.4.3 Procedure

Radiography shall be carried out according to a written procedure based on applicable codes and standards and also shall consider the safety and health aspects of the personnel engaged in radiography, as well as the inhabitants within the vicinity.

11.4.4 Radiographs

Radiographs shall be presented for review along with proper reports for the same. It shall indicate all salient aspects of the technique and invariably shall contain details required for calculating the geometrical unsharpness of radiographs, which again is one of the acceptance criteria regarding technique.

11.4.5 Acceptability Norms

Radiographs shall be interpreted and evaluated for acceptance or rejection as per UW 51(b) of ASME Section VIII Div (1) (refer to API 650, Clause 8.1.5). Refer to Annexure D for excerpts from ASME Section VIII Div (1) for quick reference.

11.4.6 Progressive or Penalty Radiography

When a section of weld seen in a radiograph is found unacceptable because of defects present in the weld, two spots adjacent to this section shall be radiographed. In case the original radiograph shows a defect-free weld of at least 75 mm toward one end of the spot, an additional spot need not be taken on that side of the weld. If the weld on either of the adjacent sections radiographed fails to comply with requirements, additional spots shall be examined until the limits of unacceptable welding are determined. The fabricator is provided with the option to replace the entire weld done by that particular welder without resorting to further radiography. If welding is replaced, the inspector can ask for spot radiography on any other weld by a new welder. If any of these additional spots fail to comply with requirements, the limit of unacceptability needs to be determined, as mentioned previously.

11.4.7 Repair of Defective Welds

Removal of any defective weld is carried out either by gouging or by grinding or by a combination of both. Depending on the relative positioning of the defect with respect to the thickness of the weld, the repair is carried out from either the inside or the outside. Use of ultrasonic flaw detection is extremely useful, especially with thick shell tanks, where repair is costly by way of lost time and effort, apart from the cost of consumables and labor. Repairs need to be carried out only on the weld area where the defect is present, and hence the defect location shall be marked on the weld as accurately as possible so as to avoid repair of any good weld. Keep in mind when marking repairs that the chance of having a defect in the repaired weld is higher compared to that of an original weld, as repairs are carried out under more hostile conditions. The repaired radiographic segment shall undergo radiography and shall meet acceptance criteria. This repaired radiograph shall be reviewed along with the original radiograph to ensure that the new radiography was carried out at the required location and also to ensure that the defect was removed satisfactorily.

11.4.8 Records

The radiographs and their reports and weld map showing the spots radiographed are the records pertaining to radiography. As a standard practice, all records shall be handed over to the owner of the facility as a part of the manufacturer's record book on completion of work. However, a copy of the same shall be available with the fabricator at least for a period of 5 years or more as desired by the fabricator.

11.4.9 Specific Requirements for Radiography from Some Clients in the Oil and Gas Industry

- The procedure mentioned in Section 11.4.3 shall be accompanied by specific technique sheets covering all types of joints envisaged in radiography as per scope.

- Interpretation of radiographs on behalf of the contractor shall be by ASNT SNT-TC-1A Level II or Ill qualified personnel or with equivalent qualification as agreed.
- All volumetric NDT shall be by radiography except for lamination checks or as agreed between contractor and client.
- X-radiography shall be used for all hydrocarbon storage tanks unless otherwise agreed for the use of the Iridium 192 isotope.
- Radiographic examination essentially shall follow work progression. Where shell weld seams require 100% examination, this shall be delayed at least 24 hours after completion of welding.
- Locations for the spot radiography of horizontal and vertical seams and intersections shall be selected at random by the client inspector.
- Identification markers shall appear on the films as radiographic images.
- When additional radiography adjacent to a defective area of a spot or randomly radiographed vertical seam reveals further unacceptable defects, then that vertical seam shall be subject to 100% radiographic examination.
- Wire-type image quality indicators are preferred and shall be available on each film, in independent single exposures. Radiographic sensitivity shall be 2% or better.
- The density through the weld metal shall not be less than 2.0 and not more than 3.2. The density in the parent metal adjacent to the weld shall not exceed 4.0. Film density shall be assessed using a suitably calibrated densitometer. Viewing facilities shall be capable of reading films with these densities.
- The type of film shall be ASTM A 1815 Class II (high contrast, low graininess; e.g., Kodak Industrex AA 400, Agfa D7, Fuji IX 100, or equivalent).
- Intensifying salt screens shall not be used.
- Exposure conditions shall be attested in the report by the subcontractor responsible for carrying out radiography.
- After interpretation and evaluation by ASNT Level II or III qualified personnel, radiographs complete with reports are required to be reviewed and accepted by the client.
- After completion and acceptance of a tank, the radiographic films including those showing defects to be repaired shall be handed over to the client and properly sorted, boxed, and indexed.

11.5 Ultrasonic Testing

As per API 650, ultrasonic testing (UT) is not mandatory. However, on agreement with the purchaser, inspection of weld using the ultrasonic test method is permitted in lieu of radiography. When ultrasonic testing is carried out in lieu of radiography as permitted, examinations need to be carried out as per Appendix U in API 650, which gives norms for acceptance of defects as well.

In addition to this, on many occasions, client specifications require ultrasonic testing as an additional requirement. Use of the ultrasonic test method is very helpful when shell thicknesses are comparatively higher. Furthermore, ultrasonic testing is very effective and foolproof in evaluating the corner welds such as the weld between nozzles and manholes and the shell. Because of this, it is always better to have ultrasonic testing procedures in position prior to the start of fabrication. The procedure shall address all types of weld configurations that may arise in its purview with other salient parameters involved in scanning and shall be approved by the client or consultant. Furthermore, examiners who perform ultrasonic testing shall be qualified to Level II or III of ASNT, as in the case of radiography.

When ultrasonic testing (not in lieu of radiography) is called for in the specification, it shall be carried out as per Article 5 of ASME Section V. Since this ultrasonic testing (additional) is not mandatory, acceptance criteria for the same may need to be agreed upon by the purchaser and manufacturer preferably before the start of work.

11.6 Magnetic Particle Testing

When magnetic particle testing (MPT) is specified in the order, it shall be carried out as per Article 7 of ASME Section V. A procedure covering testing of all anticipated types of welds shall be prepared with a proper reporting format indicating all salient parameters of the test and is to be approved by the client or consultant. Personnel performing MPT shall be qualified to ASNT SNT-TC-1A Level II or III. The vision requirements specified in API 650 with regard to NDT technicians are similar to the requirements of SNT-TC-1A, and hence API 650, through Clause 8.2.3, puts no additional requirement forward.

Criteria for acceptance, removal, and repair of defects revealed in MPT shall be in accordance with ASME Section VIII Div (1) Appendix 6, paragraphs 6-3, 6-4, and 6-5.

11.7 Liquid Penetrant Testing

When liquid penetrant testing (LPT) is specified in the order, it shall be carried out as per Article 6 of ASME Section V. As in the case of other NDT methods, a written procedure covering testing of all anticipated types of welds shall be prepared with proper reporting format indicating all salient parameters of the test and is to be approved by the client/consultant. Personnel performing and evaluating LPT shall be qualified to ASNT SNT-TC-1A Level II or III. Here also, the vision requirements specified in API 650 with regard to NDT technicians are similar to requirements of SNT-TC-1A and hence no additional requirement is put forward by API 650 through Clauses 8.4.3

Criteria for acceptance, removal and repair of defects revealed in LPT shall be in accordance with ASME Section VIII Div (1) Appendix 8, paragraphs 8-3, 8-4, and 8-5.

11.8 Visual Examination or Testing

As indicated in the inspection and test summary in Chapter 10, every weld of a storage tank is expected to be inspected visually and cleared off personally by one of the inspectors according to the norm specified in API 650, as well as in client specifications. In spite of all these written requirements, on many occasions, obvious and glaring defects can be noticed at the time of hydrostatic testing and subsequent surface cleaning and coating and lining application of the storage tanks. Though it might be unintentional and due to human errors, most of the time it is felt that this results from a lack of clear strategy to clear each and every weld of a storage tank through visual examination or visual testing (VT). Therefore, it is considered important to evolve a clear strategy to carry out visual inspection of storage tanks as described next.

11.8.1 Visual Examination Strategy

In this regard, the right method is to list out all welds that are coning on the storage tank. Once this is done, it shall be ensured with the use of a checklist that each and every weld has undergone visual inspection. To not consume more time for visual inspection of welds at the final stage, the following strategy to visually clear welds progressively at a convenient stage of manufacture would stagger the burden of visual inspection considerably over the entire manufacturing cycle. Though it is advisable to carry out the final visual inspection of the surface of the tank after completion of all welding (including capacitor discharge welding of insulation clips and similar items), in practice this might not be possible because of the short time span available to carry out VT in one shot. Therefore, as a via media stagewise VT clearing methodology is proposed as below.

Sl. No.	Component Weld Description	Proposed Last Stage to Clear VT of Typical Welds
Bottom Plates		
1	VT of annular plate butt welds	Before random radiographic testing (RT) of radial joints
2	Bottom plate fillet welds	Before vacuum box testing or magnetic particle testing (MPT)
Shell		
3	Shell vertical joints	After completion of weld from both sides before release for RT
4	Shell horizontal joints	After completion of weld from both sides before release for RT
5	Clean-out doors and other flush-type joint welds	After welding before pad air test or RT as applicable
6	Welds of nozzles and manways with pads	Before pad air test
7	Welds of nozzles and manways without pads	Before MPT or hydrostatic test as applicable
8	Other welded attachments on shell	Progressive with completion (both inside and outsides) and to be signed off immediately on clearing each of the attachments in the checklist prior to hydrostatic test
9	Erection cleats and temporary attachment removed areas	Before MPT or LPT as applicable
10	Plate surface (inside and outside)	Arc strikes and handling damages before hydrostatic test

11.8.2 Vision Requirement for Visual Inspectors

All inspectors engaged in visual examination shall have vision (with correction, if necessary) to be able to read a Jaeger Type 2 standard chart at a distance of not less than 300 mm (12″) and is capable of passing a color contrast test and shall be reaffirmed annually. The vision requirement in API 650 is slightly more stringent than that recommended in SNT-TC-1A of ASNT and hence are elaborated in Clause 8.5.1 of API 650. Though vision is a primary requirement, the inspector's competency in work is also an equally or more important aspect in qualifying a person as visual inspector.

11.8.3 Acceptance Criteria for Visual Examination

Crater cracks, other surface cracks, or arc strikes in or adjacent to welded joints are not permitted. In addition, limits are set for undercuts of various types of welds.

The maximum permissible undercut is 0.4 mm (1/64") in depth for vertical butt joints, vertically oriented permanent attachments, attachment welds for nozzles, manholes, flush-type openings, and inside shell-to-bottom welds. For full acceptance criteria with regard to undercuts and surface pores, see Clause 8.5.2 of API 650.

Weld Reinforcement

The reinforcement limits specified for vertical joints range from 2.5 mm to 5 mm, whereas those for horizontal joints range from 3 mm to 6 mm, based on plate thickness. As this limit produces wide variation in density between the weld and the parent plate, the reinforcement limit is restricted from 1.5 mm to 3 mm for the locations designated for spot radiography. The exact reinforcement limits in as welded condition is provided in Clause 8.1.3.4, and those for radiography locations are provided in Clause 8.5.2 of API 650, respectively.

11.9 Weld Maps

Contractors shall maintain a weld map of the shell weld to show locations of random spot radiography carried out, with unique identification provided for each spot radiographed. Apart from the required spots, the weld map also shall indicate penalty spots taken because of defective welding. It is recommended to use different color codes for depicting originally selected, repair, and penalty spots on the weld map. If required, separate weld maps shall be prepared for each NDT technique adopted in the tank construction.

11.10 Documentation of NDT

Documents in support of NDT shall be reports against each and every radiograph or other NDT performed on the tank during construction as per requirements. However, with this document alone, verifying the coverage as per requirement may not be possible. Therefore, it is suggested to have a consolidated summary of NDT reports depicting report numbers against each of the spots identified in the weld map.

12

Other Tests

12.1 Vacuum Box Testing

Vacuum box testing is a practical technique of testing in comparison to quantitative measures used to examine objects. Vacuum box testing can be used to test objects on which a pressure differential can be created across the area to be examined. Common application areas of vacuum box testing include piping systems, pressure vessels, and storage tanks. There are various types of boxes or frames used in vacuum box testing depending on the application area and its configuration. Vacuum box inspection can be carried out on lap welds, butt welds, and fillet welds. These vacuum boxes are used to examine a small and specific portion of a welded area to produce accurate results. Vacuum boxes used in vacuum box testing are fabricated of thick, clear Perspex with a closed cell foam seal along the bottom edge. A vacuum gauge and hose coupling are fitted in the box before the testing begins. Then a soapy solution is put on the line of the welding of the object under test. After that, the vacuum box is placed over the testing object, and a vacuity is created inside the box. The object is then observed for either a drop in vacuum or a bubble formation on the surface, in which case the weld is considered defective.

Vacuum box testing is a very popular leak-testing technique that is widely used to test various objects against defects, flaws, leaks, or other imperfections, being an efficient and effective method to test welded objects. There are basically two main types of vacuum box systems used for performing vacuum testing:

- *High-pressure models* are mainly used for inspection of above ground storage tanks that have been in service.
- *Low-pressure models* are used for weld integrity checks on new tank builds where gross defects are the main concern.

In storage tanks, vacuum box testing is extensively used to check for through defects in the welding of bottom and annular plates of the storage tank.

12.1.1 Vacuum Box Test Procedure

Vacuum box testing on tank welds shall be carried out based on a written procedure, duly approved by all concerned in accordance with code and contract specifications, and it shall essentially contain the following details:

1. List of reference codes and specifications
2. Type of equipment proposed
3. Range of pressure gauge and its calibration
4. Details of bubble-forming solution
5. Duration of test
6. Overlap required during test
7. Surface preparations and prior inspection requirements
8. Temperature and light requirements on test surface
9. Personnel qualification
10. Evaluation criteria
11. Repair and retests
12. Inspection report format

12.1.2 Vacuum Box Testing (Standard and Client Requirements)

Sl. No.	Activity Description	Requirements	
		Standard Requirement (API 650)	Client Requirement
1	Written procedure	Required	
2	Visual examination before vacuum box testing	Required	
3	Vacuum required	21 kPa (3 lbf/in^2 or 6 in. mercury) to 35 kPa (5 lbf/in^2 or 10 in. mercury)	Other pressures: 56 kPa (8 lbf/in^2 or16 in. mercury) to 70 kPa (10 lbf/in^2 or 20 in. mercury)
4	Vision check for operator	Required annually	
5	Competency in technique	Required	
6	Overlap required	50 mm with previously viewed section	
7	Temperature of surface	Between 4°C (40°F) and 52°C (125°F)	
8	Light intensity	1,000 lux at the point of examination	

Sl. No.	Activity Description	Requirements	
		Standard Requirement (API 650)	Client Requirement
9	Duration	Duration of 5 seconds or time required to view the area, whichever is greater	
10	Report	Required	
11	Pressure gauge calibration		Valid calibration required at time of testing required
12	Weld map with weld details		Required
13	Posttest cleaning	Required; suitable for subsequent operation	
14	Rejection criteria	Through-thickness leak indicated by continuous formation or growth of a bubble(s) or foam, produced by air passing through the thickness Large opening leak, indicated by a quick bursting bubble or spitting response at the initial setting of vacuum box	

12.1.3 Equipment

Vacuum box testing is performed by using a box with a visible window of fiberglass (i.e., 150 mm [6″] wide by 750 mm [30″] long metallic box with a fiberglass window). The open bottom is sealed against the tank surface during the test by a sponge rubber gasket. The test scheme shall have suitable connections, necessary valves, and calibrated vacuum gauges in the range of 0 kPa to 100 kPa (0 psi to 15 psi) or equivalent.

The test scheme shall be demonstrated with a sample test block by application of the bubble solution at the site before the test is conducted. The bubble-forming solution shall produce a film that does not break away from the area tested, and bubbles formed shall not break rapidly because of air drying or low surface tension. Soaps or detergents designed specifically for cleaning shall not be used for the bubble-forming solution.

A vacuum can be drawn on the box by any convenient method, such as connection to a gasoline or diesel motor intake manifold or to an air ejector or special vacuum pump. The gauge shall register a partial vacuum of at least 15 kPa (2 psi) below atmospheric pressure.

The bubble-forming solution brand name or type shall be known prior to execution and be included in the procedure mentioned in Section 12.1.1.

12.1.4 Types of Vacuum Boxes

	Type of Vacuum Box	Use
Flat bottom		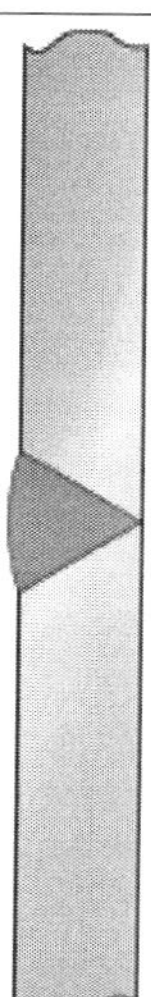 This is a flat bottom box for testing butt welds on flat surfaces.
Flat bottom		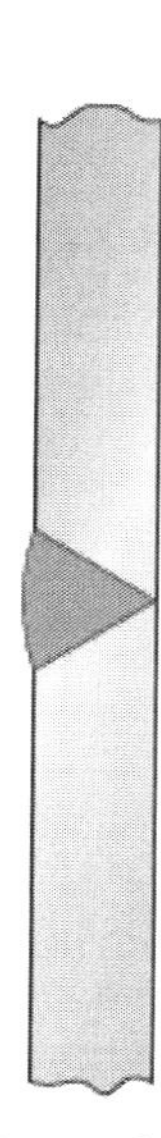 This is a flat bottom box for testing butt welds on flat surfaces (large surface area).

Lap joint

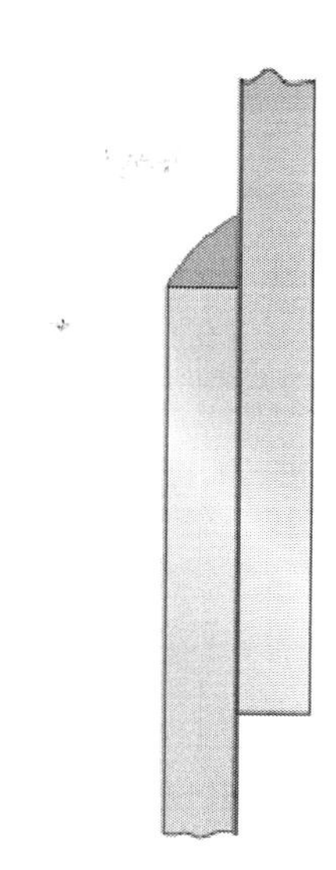

This is a flat bottom box for testing lap joint welds of 3/8″ plate on flat surfaces.

Corner

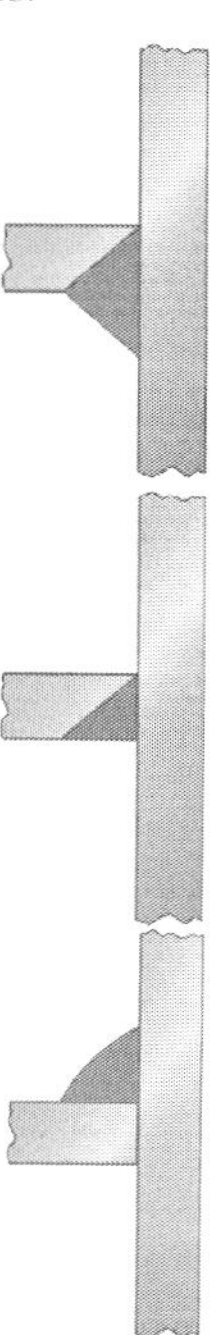

Corner vacuum boxes are specifically designed for testing the inside corner, where the bottom meets the sidewall at 90 degrees.

Corner (for small diameters)

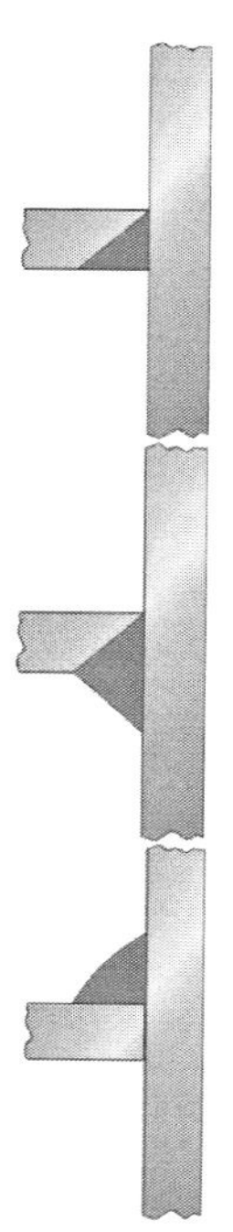

Corner vacuum boxes are specifically designed for testing the inside corner, where the bottom meets the sidewall at 90 degrees (for smaller diameter).

Corner (three corner inside)

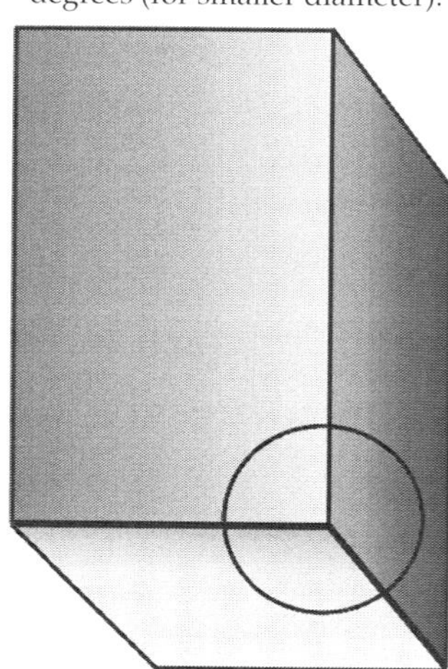

The vacuum box is specifically designed for testing the inside corner, where the bottom meets two vertical walls at 90 degrees.

Corner (outside)

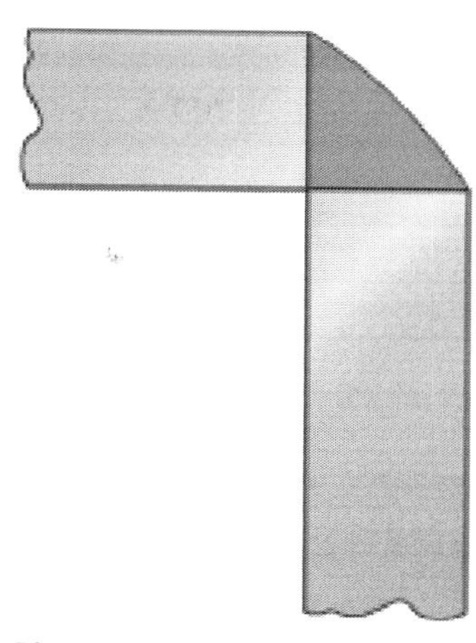

The vacuum box is specifically designed for testing the outside 90 degree corner.

Curved

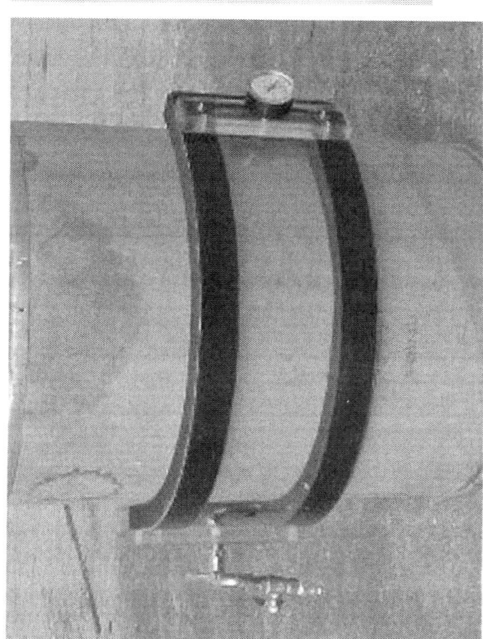

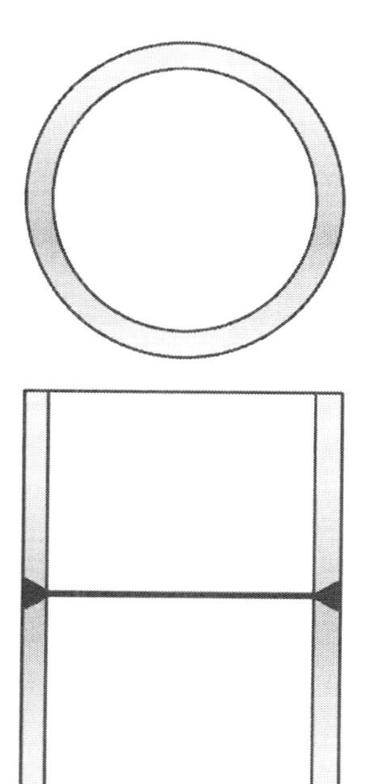

The vacuum box is designed for groove welds in pipes.

Flex box for testing welds on a radius

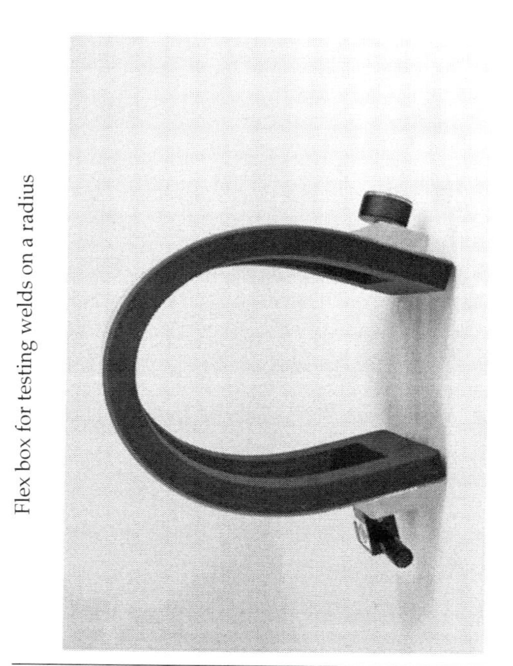

The vacuum box is designed to test a convex or concave radius. The surface must be curved. This type of box may not function properly on a flat surface.

Note: Picture courtesy M/s Tech Service Products Inc.

Vacuum box testing is carried out on all bottom plate lap welding and also corner welds and sometimes roof plate lap welds as well. This is carried out using a vacuum box of approximately 150 mm (6″) width and 750 mm (30″) length, with a clear a window at top, which provides a proper view of the area under inspection. During testing, there shall be adequate lighting to facilitate proper evaluation and interpretation of the test. The underside (the side that comes in contact with the bottom plate) of the box shall be held against the bottom plate of the tank using suitable gaskets so that the applied vacuum is maintained during testing. The box shall be provided with necessary accessories like valves and gauges as required in applicable code and specifications. The film solution shall be used to detect leaks while the vacuum is maintained inside the box.

12.1.5 Methodology of Testing

As mentioned earlier in Section 12.1.1, vacuum box testing shall be carried out according to a written procedure duly approved by all parties involved in the contract, mainly the consultant and the client. In addition to the requirements spelled out in Section 12.1.1, the procedure shall also address the following aspects related to the actual performance of the test at the site:

1. Visual examination of the weld joints is made prior to conducting the vacuum box test.
2. Physical verification of the vacuum pump and its gaskets is made to assess whether they are capable of producing satisfactory results.
3. The adequacy of the wetness of the film solution (when applied to a dry area) to generate bubbles is ensured.
4. The vacuum shall be as indicated in the table provided in Section 12.1.2 and shall be monitored through calibrated pressure gauges.
5. The operator shall be qualified to carry out the test as specified in the written procedure and shall undergo a vision test annually.
6. The operator shall ensure a minimum overlap between consecutive tests as specified in code and specifications.
7. The test shall not be performed when the light intensity available is below that specified. Furthermore, the duration of the test also shall be maintained as required.

12.1.6 Acceptance Criteria

Through-thickness leaks of any sort are treated as unacceptable in API 650. The leaks are classified under the following two categories:

1. A through-thickness leak is indicated by continuous formation or growth of a bubble(s) or foam, produced by air passing through the thickness.
2. A large opening leak is indicated by a quick bursting bubble or spitting response at the initial setting of the vacuum box.

Both the leaks are unacceptable and shall be repaired and tested again.

12.1.7 Records

A report of the test indicating all salient aspects mentioned previously shall be a part of the final documentation pertaining to the tank, along with details of welds covered by the report and a weld map.

Though API 650 permits leak testing using a tracer gas and suitable detector in lieu of vacuum box testing, the same is not addressed in this book, as this is intended to provide simple, effective, and low-cost options that are in strict compliance to API 650 requirements. However, if client specification requires the tracer gas leak detection system, details pertaining to the same can be had from Clause 8.6.1 of API 650.

12.2 Pneumatic Testing of Reinforcement Pads

All reinforcement pads provided as compensation against openings made on the tank shall be subjected to pneumatic testing before the final hydrostatic testing of the tank. Reinforcement pads are required for nozzles of 65 NB (3″) and above. Reinforcement pads so provided shall have telltale holes, intended to facilitate the escape of hot gases during welding, to detect any leaks during hydrostatic testing, and to detect leaks from nozzle or manway welds during service, which can be used for pneumatic testing of pads as well.

12.2.1 Requirements

As mentioned previously, all reinforcement pads provided as compensation against openings made on the tank shall be subjected to pneumatic testing before the final hydrostatic testing of the tank. To facilitate this, all reinforcement pads shall be provided with at least one telltale hole in each segment of the pad, if they are made in segments, and these holes shall be open to the atmosphere even during the service life of the tank. Telltale holes shall be of 6 mm (1/4″) in diameter and internally threaded so as to hold the pneumatic test equipment. The ideal location to place this telltale hole shall be at 25 mm (1″) from the periphery of the pad at the horizontal centerline. (For details refer to Figure 5.8 of API 650.)

12.2.2 Test Pressure and Methodology

Testing of reinforcement pads shall be carried out at a pressure of 100 kPa (15 psig) gauge. Pressure is applied using an attachment similar to that shown in Figure 12.1 between the pad plate and the shell. The film solution shall be

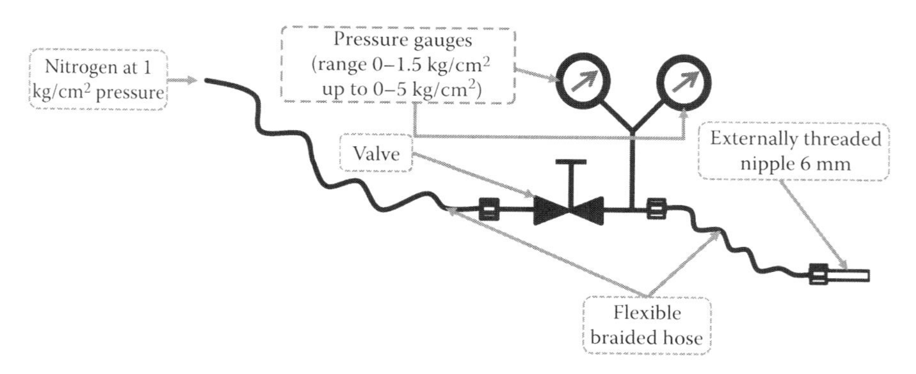

FIGURE 12.1
Arrangement for reinforcement pad testing.

applied to all welds between the pad, manhole/nozzle, and shell, from both inside and outside the tank. The welds that are to be tested using the film solution are indicated in Figure 12.2.

12.2.3 Arrangements

1	Shell to nozzle weld	Covered by pad air test
2	Pad to nozzle (FP) weld	N/A
3	Pad to nozzle (fillet)	Covered by pad air test
4	Pad to shell (fillet)	Covered by pad air test
5	Telltale hole	
6	Nozzle/manway neck	
7	Reinforcement pad	
8	Shell	

Pads provided on the shell for purposes other than compensation in lieu of an opening made on the shell or roof need not be subjected to this test. However, such pads also shall be given telltale holes as provided in the case of reinforcement pads given as compensation to openings made on the shell or roof. Because of this, telltale holes on such pads need not be tapped. During a rainy season, accumulation of water in the gap between the pad and the shell or roof is a usual phenomenon, which may give rise to corrosion, especially in the acidic atmosphere usually present in the chemical plant environment. To ward off this issue, all threaded or unthreaded telltale holes shall be plugged using heavy grease. As grease gets hardened because of weather conditions, it may be replaced when found ineffective.

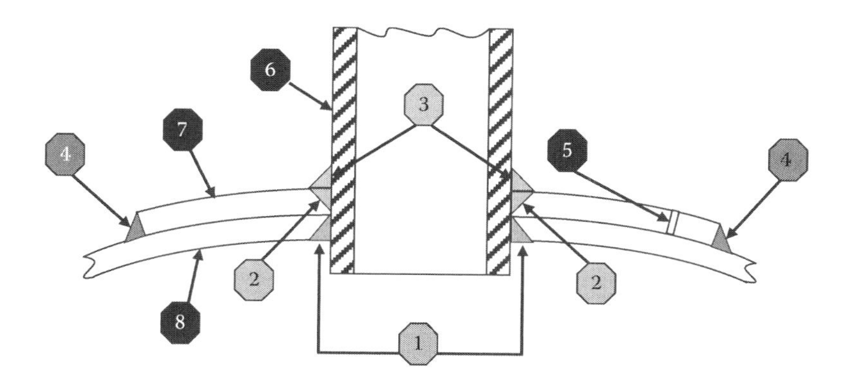

FIGURE 12.2
Welds to be inspected.

Photograph of pad air test of a typical shop fabricated tank.

The test needs to be carried out according to a written procedure duly approved by all concerned. In case magnetic particle testing or liquid penetrant testing of nozzle or pad weld is required either by code or by specification from the client, this shall be completed prior to the pneumatic testing of the pads.

1. As mentioned earlier, this pneumatic test shall be performed prior to heat treatment (if applicable) and a hydrostatic test.
2. Attachment as per Figure 12.1A containing two pressure gauges and an isolation valve shall be connected to a 6 mm (1/4″) internally threaded telltale hole, and air pressure shall be applied gradually. Pressure gauges shall have valid calibration at the time of the test. The range of pressure gauges used shall be within 1.5 to 4 times the test pressure.
3. Open the inlet valve until the desired pressure is reached, 100 kPa (15 psig) with either dry air or nitrogen. Maintain that pressure until the test is completed.
4. While the pad is pressurized, apply a nonchloride film solution over the reinforcing pad welds on the inside and outside of the tank using a hand spray pump or brush. Welds that are to be inspected using a film solution are shown in Figure 12.1B.
5. Other factors to be considered during the test are as follows:
   - Temperature limits of the metal surface during the test shall be between 4°C and 52°C (40°F and 125°F), unless the film solution is proven to work at temperatures outside these limits, either by testing or by the manufacturer's recommendations.
   - A minimum light intensity of 1,000 lux at the point of examination is required during the test and evaluation for leaks.
   - The examiner shall have a valid vision certification.

6. Observe for any sign of leaks, the absence of which confirms acceptance, and pressure can be brought back to atmospheric. Otherwise, depressurize, repair, and retest in the same manner.
7. When the pad has successfully passed the test, documentation to this effect shall be made and countersigned by all parties involved. This shall be included in the manufacturer's record book.

12.3 Hydrostatic Testing of Tank

12.3.1 Testing of Tank Shell

API 650 proposes one of the following to assess the integrity of the shell of a storage tank. This test is to be carried out after the entire construction work of the tank is completed, including the roof and all other connected structures. However, tanks designed according to Appendix F of API 650 is excluded from the scope of this testing.

If water is available for carrying out the hydrostatic test, the tank shall be filled with water

1. up to a maximum design liquid level,
2. up to 50 mm (2″) above the weld connecting the roof plate or compression bar to the top angle or shell in the case of tanks with a tight roof, or
3. to a level lower than that just specified, when the maximum level is limited by overflow nozzles.

12.3.2 Standard and Client Requirements for Hydrostatic Testing

Sl. No.	Activity Description	Requirements	
		Standard Requirement (API 650)	Client Requirement
1	Written procedure		Required and shall address requirements in detail
2	Stage	Before connecting external piping After completion of all welding	
3	Coverage	Shell welds All welds above water level shall be tested either pneumatically or using chalk and oil	Pneumatic or vacuum box testing

(Continued)

Sl. No.	Activity Description	Requirements	
		Standard Requirement (API 650)	Client Requirement
4	Medium	Potable water preferred	Quality of water and additives if other types of water such as well water, seawater, etc. are proposed
5	Filling rate	Restrictions as per Clause 7.3.6.5 of API 650	Restricted further by clients based on the type of foundation selected
6	Settlement measurement	Required	
7	Duration of hydrostatic test	24 hours Water to be drained off within 14 days if not proper additives required	48 hours Requirements for additives
8	Welds above water fill height	Pneumatic test or chalk oil test	Pneumatic test
9	Temperature of surface	Shall not be colder than minimum design metal temperature	
10	Testing of roof	Required	
11	Roof test pressure	Not exceeding weight of roof plate	
12	Report	Required	
13	Pressure gauge calibration		Required if applicable; manometer against calibrated steel rule preferred
14	Weld map with weld details		Required as checklist
15	Posttest cleaning		Required; suitable for subsequent operation
16	Rejection criteria	Leaks	Leaks or sweating

12.3.3 Recommended Contents for the Hydrostatic Test Procedure

Sl. No.	Description of Requirement
1	Stage at which the test is carried out
2	Medium proposed (potable water preferred for carbon steel tanks)
3	In case of brackish water, dosing requirements
4	Duration of retention expected including filling, inspection, and dewatering
5	Schematic of filling arrangements with dosing
6	Measurement of height of filling
7	Settlement measurement (requirements and methodology), preferably as a stand-alone document
8	Disposal of test water

12.3.4 Water Filling and Draining Rates

Sl. No.	Bottom Course Thickness	Tank Portion	Maximum Filling and Draining Rate
1	Less than 22 mm (7/8″)	Top course	300 mm (12″)/hour
2		Below top course	460 mm (18″)/hour
3	22 mm (7/8″) and thicker	Top third of tank	230 mm (9″)/hour
4		Middle third of tank	300 mm (12″)/hour
5		Bottom third of tank	460 mm (18″)/hour

The tank shall be inspected frequently for any leaks during the filling operation. All welded joints above the test water level shall be examined as mentioned in Section 12.6 if no other test is required for those welds. Apart from the inspection of welds at various stages of filling, settlement of the tank also shall be recorded, the minimum requirements for which are provided in the following Section 12.4. Since API 650 is silent on the acceptance criteria for this aspect, which is one of the essential criteria for the evaluation of the tank during in-service inspection while in service, the acceptance criteria given in API 653 and has been utilized for the purpose and is given in ensuing sections.

12.3.5 Filling of Test Medium

Potable water is the preferred medium for hydrostatic testing of storage tanks made of carbon steel. However, testing with other types of water, like brackish water, is not prohibited on account of the availability of potable water in huge quantities at the site. In such cases the dosing requirement shall be established considering the quality of test water available at the site, duly taking into account the probable detention time of water inside the tank until full draining and drying of the tank.

The tank shall be filled with water by suitable low-pressure filling lines, while venting through the uppermost nozzles. The capacity of the filling pump shall be controlled to maintain adequate venting to avoid air entrapments. Before applying pressure, the test operator shall check all connected equipment to ensure that all low-pressure filling lines and other appurtenances that should not be subjected to test pressure have been positively disconnected or isolated by valves or other suitable means.

1. For a site-fabricated tank, the water-filling height shall be restricted to the maximum liquid level or up to the curb angle.
2. The water filling shall be carried out in four stages (25%, 50%, 75%, and 100% of the maximum liquid level of the tank).

3. Any leaks observed in the shell joints during the hydrostatic test shall be noted and repaired with the water level lowered down to a minimum of 300 mm below the leak spot.
4. Repairs shall be carried out as per an approved procedure and shall be subject to applicable nondestructive testing (NDT). Upon completion of these two activities, the tank test shall be resumed.
5. After each stage of filling, a load stabilization period of 24 hours shall be observed between each stage of filling.
6. During the stabilization period, the settlement reading shall be recorded jointly with the contractor, consultant, and client representatives as required using a dumpy level or a theodolite.
7. In addition to the filling rate indicated in Section 12.3.1, some clients specify a per-day cap on the filling rate as 5 m of height of water per day.
8. After completion of water filling, the full-height water load shall be maintained for 24 hours. After 24 hours the final settlement reading shall be recorded jointly.

In addition to tank settlement measurements taken at one-fourth, one-half, three-quarters, and full height, they shall be taken again after emptying the tank after the hydrostatic test.

12.3.6 Visual Inspection

During the test, inspection for leakage, bulging, or other visible defects shall be made on the whole body of the tank. Close examination shall be carried out on all weld joints and connections. The inspection shall be performed by experienced and qualified personnel from the contractor, consultant, or client side as required.

12.3.7 Safety Precautions

Warning signs shall be displayed at roped-off areas indicating that a test is in progress. Only authorized personnel shall be allowed within the test area during testing. All authorized personnel allowed within the test area shall stay clear of the tank during testing.

12.3.8 Repair and Retest

During inspection, if leaks, cracks, or any other defects are observed, the same shall be repaired as per the approved procedure and by qualified welders. After successful repair and NDT, the tank shall be retested as per the original written procedure.

If during the test and inspection any excessive change in the shape of the tank is noticed, engineering verification shall be carried out and means shall be provided in the tank to retain the designated shape within permissible limits.

12.3.9 Draining

After completion of the test, a check valve shall be installed in reverse position, and the drain valve shall be opened to remove water from the tank. A pump shall be used for complete drainage from the sump. The water shall be disposed of at an appropriate place identified by the client.

After all water has been drained, the tank shall be examined internally for cleanliness and drying, and the final inspection shall be carried out.

12.3.10 Cleaning

The tank shall be emptied at a maximum water level variation rate of 5 meters per day. Care shall be taken that the top manholes are open during the emptying operation of conical roof tanks to avoid vacuum generation within the tank during this process.

If brackish or other similar water is used for hydrostatic testing, then the tank shall be thoroughly rinsed with potable water. After draining the water completely (including rinsed water), the tank shall be thoroughly cleaned, free from dirt and foreign materials, and dried by air.

12.3.11 Documentation

The documents in the following table shall accompany the hydrostatic test of a storage tank.

Sl. No.	Description
1	Hydrostatic test report in format approved covering details of items included in the test
2	Welding completion checklist
3	NDT completion checklist (with weld maps)
4	Post weld heat treatment (PWHT) completion (if applicable) checklist
5	Pad air test completion checklist
6	Vacuum box test completion checklist
7	Mechanical completion checklist
8	Settlement measurement records
9	Roof air test reports with weld maps
10	Pressure gauge and steel ruler calibration records
11	Punch list if any outstanding

12.4 Measurement of Settlement of Tank during Hydrostatic Testing

When settlement of the tank is expected (especially in the case of sand pad foundations) during hydrostatic testing, the initial level of the bottom plate projecting outside the tank shell has to be recorded. During the initial survey of the level of the bottom plate, the tank shall be empty. The number of points where settlement is to be determined is taken as equally spaced intervals around the tank circumference not exceeding 10 m (32′) or shell diameter D/10 or a minimum of 8, where D is the tank diameter in feet. If fractions come up when the number of points is being determined, they shall be rounded off to the next higher whole number.

These points shall be marked equidistantly on the periphery of the tank on the shell at the bottom (say at about 200 mm to 300 mm) above the bottom plate so as to clear possible gussets or anchor chairs. Locations thus identified shall be brought upward on the shell by about a meter from the bottom plate by placing markers at a fixed distance (say a meter) above the bottom plate. The level of these points shall be recorded with respect to an external reference prior to the start of the filling operation. A setting marker shall be welded, and an initial level reading shall be taken with respect to the permanent benchmark using the dumpy level/total station. The level instrument shall be set up at least one and a half times the tank diameter away from the tank when settlement readings are taken. A minimum of six sets of settlement readings need to be taken, as listed in the following table.

Tank Settlement Measurement

Sl. No.	Description	Remarks
1	After erection of tank prior to water filling	Initial level
2	At 1/4 full height (± 600 mm)	After stabilization for 24 hours
3	At 1/2 full height (± 600 mm)	
4	At 3/4 full height (± 600 mm)	
5	At full height (± 600 mm)	
6	At 1/2 full height (± 600mm) during emptying[a]	
7	After emptying	Final level

[a] Additional requirement by clients in the oil and gas industry.

To obtain a clear pattern of settlement, some client specifications require settlement readings taken at every 1 m of water fill (without allowing for stabilization time). This may be required when the strength of the sand pad foundation is in doubt.

In the case of new storage tanks, the types of settlement expected (as with sand pad foundations) are uniform settlement and out-of-plane settlement. Uniform settlement is not considered very critical and shall be present in almost all cases to the tune of 50 mm to 60 mm. Uneven or out-of-plane settlement observed during the hydrostatic test is to be considered detrimental, as it is expected to worsen during the service life of the tank. As mentioned earlier, API 650 is silent in this regard; acceptance criteria specified in Annex B of API 653 can be used to determine the acceptable level of out of plane settlement. Furthermore, these criteria, which are beyond the jurisdiction of API 650, need client or consultant concurrence based on an engineering assessment of the situation.

In addition, internal bottom elevation measurements (buckling of the bottom plate) are required before and after hydrostatic testing, as indicated in the following table.

Internal Bottom Elevation Measurement (before and after Hydrostatic Testing)

Sl. No.	Description of Stage	Frequency
1	Number of diametrical lines required.	Equally spaced with maximum separation at 10 m (32′) over periphery. Minimum number of lines is four.
2	Number of spots required along diametrical lines.	At 3 m (10′) interval along diametrical lines.

12.5 Alternate Tests in Lieu of Hydrostatic Testing

If water is too scarce in the vicinity to fill up the tank for the hydrostatic test, all weld joints shall be tested using highly penetrating oil. The inside surface of welds shall be painted with oil and inspected from the outside for leakages.

Another alternative is to apply a vacuum or air pressure on one side of the weld and observe the welds from the other side for any leakages. The pressure or vacuum shall be applied as discussed in the following Section 12.6. In addition, a combination of these two alternatives is also permitted as per API 650.

While alternates are specified, the hydrostatic fill test is considered the most comprehensive as far as the integrity of the tank is considered, and hence these alternatives are not often accepted by many client specifications, which insist on full hydrostatic testing.

12.6 Testing of Roof

The roof of a storage tank is tested after hydrostatic testing but before draining the water. For hydrostatic testing, water is usually filled up to the top-most permitted height as mentioned in Clause 7.3.5(1) of API 650. Welds above this level are usually tested in any of the following manner, especially roof welds:

1. Apply internal air pressure that does not exceed the weight of the roof plates, and apply a film solution to the weld joints to detect leaks. This requires the design group to work out the test pressure by considering the weight of the roof plates alone. Since this varies with the thickness of the plates, the diameter of the tank, and so on, this needs to be calculated on a case-by-case basis.
2. Perform vacuum box testing of the weld joints as mentioned in Section 12.1 to detect possible leaks in the roof welds.

If roofs are not designed as gas tight or tanks have vents at the top, only visual inspection need be carried out unless otherwise spelled out elsewhere.

12.7 Testing of Roof for Appendix F Tanks

The foregoing sections are applicable only to tanks that are designed as per API 650, and special conditions shall apply to roofs designed as per Appendix F.4.4 and F.7.6 of API 650.

Appendix F gives the design basis for tanks subject to small internal pressures as defined in it. So when the tank is designed as per this appendix, a slight change in roof-testing methodology is required, as described next.

Roof testing is to be carried out after completion of the tank in all respects. Water shall be filled up to the top angle or designed liquid level, and designed internal pressure shall be applied to the enclosed space above the water level and shall be held for a period of 15 minutes minimum. Air pressure shall be reduced to half the designed pressure, and all welded joints above the liquid level shall be checked for leaks by means of a film solution. Tank vents also shall be tested along with this test or after.

In the case of anchored tanks, this testing shall be carried out with a few additional precautions:

1. After the tank is filled with water, the shell and anchorage shall be visually inspected for tightness.
2. Air pressure of 1.25 times the designed pressure shall be applied to the tank filled with water to the designed liquid height.

3. The air pressure shall be reduced to the designed pressure, and the tank shall be checked for tightness.
4. In addition, all seams above the water level shall be tested using a film solution.
5. After satisfactory completion of the shell and roof testing, water shall be drained completely, and the tank shall be brought to atmospheric pressure. Anchorage shall be checked for tightness again in this condition.
6. The designed air pressure shall then be applied to the tank for a final check of the anchorage.

13

Cleaning, Internal Lining, and External Painting

13.1 Cleaning after Hydrostatic Test

After hydrostatic testing, tanks are drained off and cleaned of debris and dirt. If potable water is used for hydrostatic testing, the subsequent cleaning process required is comparatively simple and straightforward. If brackish water or other similar types of water is used for the hydrostatic test, then washing the internal surface with potable water might be required subsequent to the hydrostatic test, prior to drying.

After drain off, the tank is dried using dry air, thereby making it ready to receive suitable surface preparation or internal lining and external coating as required by service.

13.2 Planning for Surface Preparation and Lining or Painting

Painting or lining works (in general and for storage tanks) are usually planned in the following way. Paints or linings are selected after studying various factors that affect the durability of the coated film under service. The surface preparation method and painting or lining procedure is decided by considering the characteristics of the paint or lining system, the design of structures, and so on. In addition, all painting and lining specifications are decided based on the cost factor as well. The four fundamental considerations involved in deciding a painting or lining system are shown in the following charts.

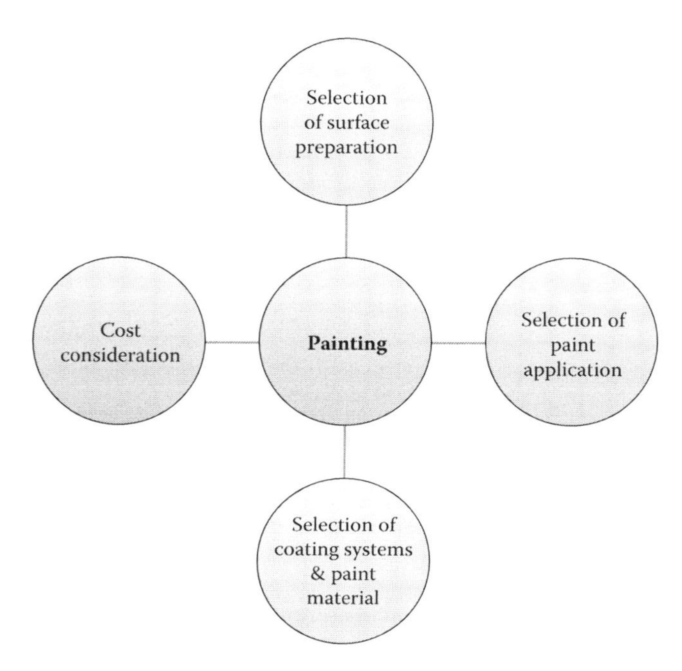

Note: Chart courtesy M/s Kansai Paints, Japan.

Further breakdown of these factors is provided in the following chart and the table following it. The purpose of this is to provide an overview of the various factors associated with the planning involved in coating works and also to impart awareness to readers about the importance of each of the aspects. Many times, the type of coating or lining system required for the tank is specified in the data sheet, and hence aspects related to cost and selection are not significant. However, for the sake of completeness, even those topics are addressed in the following chart and table to provide an overview.

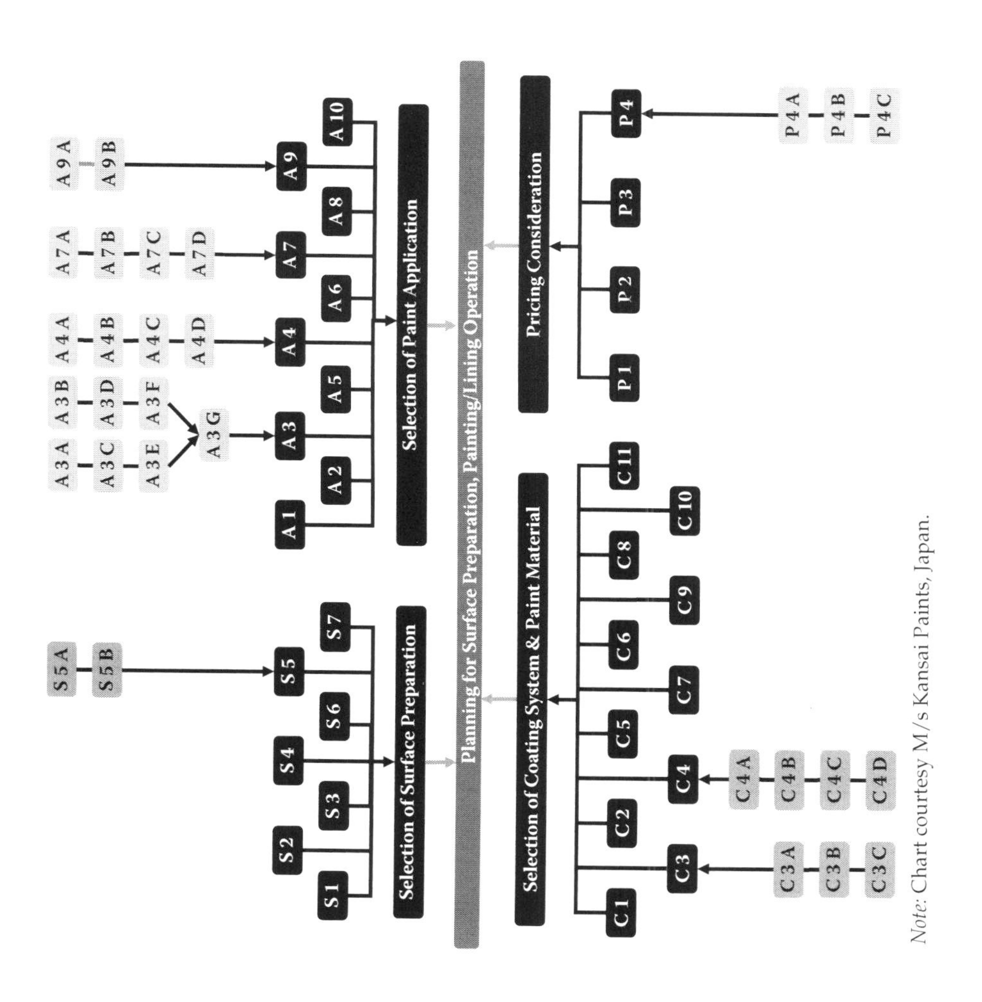

Note: Chart courtesy M/s Kansai Paints, Japan.

Planning for Surface Preparation and Painting or Lining (Decoding Table)

Sl. No.	Activity Code	Description	Sl. No.	Activity Code	Description
1		Selection of surface preparation	39		Selection of coating system and paint material
2	S 1	Surface profile	40	C 1	Exposure environment
3	S 2	Condition of surface	41	C 2	Condition of surface
4	S 3	Environmental constraints	42	C 3	Properties of paint
5	S 4	Required derusting grade	43	C 3 A	Thickness
6	S 5	Location of job	44	C 3 B	Compatibility
7	S 5 A	Postfabrication	45	C 3 C	Recoating
8	S 5 B	On-site	46	C 4	Surface encountered condition
9	S 6	Physical or chemical cleanliness	47	C 4 A	Shop primed
10	S 7	Safety	48	C 4 B	Field painted
11		Selection of paint application	49	C 4 C	Shop finish painting
12	A 1	Drying and handling	50	C 4 D	Maintenance
13	A 2	Shop painting	51	C 5	Labor considerations
14	A 3	Effect on application of paint	52	C 6	Inspection
15	A 3 A	Intercoat adhesion	53	C 7	Term of durability
16	A 3 B	Environmental constraints	54	C 8	Types of contracts, bids, and proposals
17	A 3 C	Thickness	55	C 9	Types of structures
18	A 3 D	Contact surface	56	C 10	Environmental constraints
19	A 3 E	Stripping	57	C 11	Safety
20	A 3 F	Continuity	58		Pricing considerations
21	A 3 G	Damage	59	P 1	Types of contracts, bids, and proposals
22	A 4	Application methods	60	P 2	Maintainability in future
23	A 4 A	Air spray	61	P 3	Cost factors in coating system
24	A 4 B	Brush	62	P 4	Cost factors
25	A 4 C	Roller	63	P 4 A	Materials
26	A 4 D	Airless spray	64	P 4 B	Labor
27	A 5	Safety	65	P 4 C	Equipment
28	A 6	Inspection			
29	A 7	Preapplication procedure			
30	A 7 A	Surface preparation			
31	A 7 B	Storage of paint and thinner			
32	A 7 C	Mixing and thinning			
33	A 7 D	Materials handling and use			
34	A 8	Field painting			
35	A 9	Conditions of surfaces encountered			
36	A 9 A	New construction			
37	A 9 B	Maintenance			
38	A 10	Scaffolding			

13.3 Surface Preparation (Write-up Courtesy M/s Transocean Coating, Rotterdam)

The single most important function that can influence any paint performance is the quality of surface preparation. For optimum service life, the surface shall be completely free of all contaminants that might impair performance and shall be treated as such to ensure good and permanent adhesion of the paint system proposed for the environment. The quality of surface preparation has a direct impact in the life cycle of paint and lining systems, even in the case of surface tolerant coatings, giving better performance in case surface preparation is better.

Surface preparation consists of two parts, namely, primary and secondary preparations. Primary preparation is intended to remove mill scales, rust, other corrosion products, and other foreign material from the metal surface prior to the application of shop primer.

Secondary surface preparation is intended to remove rust or any other foreign matter if any from the metal surface that has already been coated with shop primer, prior to the application of an anticorrosive system. All rust, rust scale, heavy chalk, or deteriorated coating shall be removed by a combination of solvent or detergent washing, hand or power tool cleaning, or abrasive blasting. Glossy areas of sound previous coating need not be removed but shall be mechanically abraded or brush blasted to create a surface profile that increases coating adhesion.

The extent to which a surface is made clean before coating is applied as a balance between the following:

- Expected performance of coating
- Paint manufacturers' recommendations
- Time available for the job
- Relative cost of various surface preparation methods available
- Access to area to be prepared
- Condition of the steel prior to surface preparation

In most of the cases, coatings cannot be applied under ideal conditions, especially for repair and maintenance jobs.

The quality of surface cleanliness that is achieved (or that is possible to achieve) would widely vary between an uncorroded high-quality steel plate with a tightly adherent mill scale (used in new constructions) and that of a tank in service for ten years, with poorly adherent coating, loose rust scales, and heavy pitting.

Any substance that prevents a coating from adhering directly to steel can be considered as a contaminant. Major contaminants for new constructions include the following:

- Moisture or water
- Oil and grease
- Ionic species from nearby sea and industrial areas
- White rust (zinc salts from weathered zinc silicate shop primers)
- Weld spatter
- Weld fume
- Cutting fume
- Burn through from welding on reverse side of the plate
- Dust and dirt from yard site and from neighboring industrial processes

In addition, for maintenance and repair situations, the presence of pitting, corrosion products, cathodic protection products, aged coatings, trapped cargoes (liquid sediments), and so on must also be considered, particularly if only localized surface preparation of severely affected areas is being carried out prior to recoating.

13.3.1 High-Pressure Freshwater Cleaning

Freshwater cleaning is necessary to remove salts, fouling, or any loose paint or other contaminants. A water pressure to the tune of 500 kg/cm^2 (approximately 7,000 psi) is typically used for removal of surface contaminants and fouling organisms such as algae.

13.3.2 Solvent Cleaning

Prior to the use of any method of surface preparation, it is essential to remove all soluble salts, oil, grease, drilling or cutting compounds, or any other surface contaminants on the surface to be coated. Perhaps the most common method adopted is solvent washing followed by wiping the surface dry with clean rags. If wipe cleaning is not carried out properly, solvent washing may spread the contamination over a larger area than the original, and hence it is very important. Proprietary emulsions, degreasing compounds, and steam cleaning are also used extensively in the industry. ISO 8504 and SSPC SP1 describe acceptable procedures for solvent cleaning.

13.3.3 Hand Tool Cleaning

This method is the slowest and least satisfactory method of surface preparation. It is frequently used in confined areas where power tool access is not possible. Scrapers, chipping hammers, and chisels can be used to remove loose, nonadherent paint, rust, or scale, but it is a laborious method and very difficult to achieve a good standard of surface preparation. Wire brushing can make the surface worse by polishing rather than cleaning a rusted surface. Soluble salts, dirt, and other contaminants are frequently trapped and coated over, leading to early paint breakdown. These methods are incomplete, since such cleaning always leaves a layer of tightly adhering rust on a steel surface. Methods for hand tool cleaning are described in SSPC SP2 and ISO 8501-1 Grade St2-B, C, or D. After cleaning, the surface is brushed, swept, dusted, and blown off with compressed air to remove all loose matter.

13.3.4 Pickling

The acid pickling process can be used for preparation of small items before coating. Items such as pipes are alkali cleaned, then washed, and later passed through an acid pickling in a bath to remove rust. This is to be followed by a thorough washing to remove all remaining acid on the surface, particularly if the item is to be painted subsequently.

13.3.5 Power Tool Cleaning

The effectiveness of cleaning using power tools rather than abrasive or water blasting methods shall depend on the effort and endurance of the operator, as working above shoulder height is especially tiring.

Typical examples of mechanical power tool cleaning are rotary wire brushes, sanding discs, and needle guns. Power tool cleaning is in general more effective and less laborious than hand tool cleaning for removal of loosely adhering mill scale, paint, and rust. However, power tool cleaning will not remove tightly adhering rust and mill scales.

Care shall be taken, especially with power brushes, not to polish a metal surface, as this could reduce the key for subsequent paint coating.

Preparation grades with power tool cleaning are specified in ISO 8501-1 and relevant preparation grades St2-B, C, or D and St3-B, C, or D. SSPC SP 11 describes various degrees of surface profiles that can be achieved through power tool cleaning.

As mentioned, some of the more popular methods falling under this category are briefly described as follows.

Rotary Power Disking

This one is the most commonly used surface preparation method in a majority of maintenance situations. It is also widely used in new constructions for preparation of welds and cut edges prior to painting. Normally silicon carbide discs are used, the grade of which is selected to suit the conditions of the surface to be abraded. It is important to change discs at regular intervals in order to maintain efficiency.

Care shall be exercised in the selection of grit size and type of disc to be utilized, so that the surface is not excessively smoothed, thereby reducing the ability of paint to adhere. Irregular and pitted surfaces may require a combination of various power tool cleaning methods to maximize effectiveness.

Power disc preparation is also widely used in new construction to clean edges of welds and intricate locations where other methods cannot reach effectively.

Mechanical Descaling

Needle guns, Roto-Peen, and other pounding-type tools are effective to some degree in removing thick rust and scale and are frequently used in maintenance works of storage tanks in service. The action of these types of devices is dependent on the cutting blade or point pounding of the surface concerned and breaking away of the scales present. This type of cleaning is effective only at the actual points of contact. The intermediate or adjacent areas are only partially cleaned, because brittle scale disintegrates, but the lowermost layer of rust and scale remains attached to the substrate, thereby resulting in reduced effectiveness of the applied coating.

Rotary Wire Brushing

This method has some merits, depending on the condition of the surface requiring treatment. Loose "powdery" rust can be removed, but hard scale will resist abrasion of wire bristles. When rust scale is intact and strongly adhered to the substrate, rotary wire brushing tends to merely burnish or polish the surface of rust scale but does not remove it. In fact, the burnished surface may give a false appearance of a well-cleaned surface, which is often misleading.

13.3.6 Blast Cleaning

This is the most commonly used surface preparation method in the industry for the application of paint. When properly carried out, abrasive blasting removes old paint, rust, salts, fouling, and so on and provides a good mechanical key (blast profile) for the new coating. However, after abrasive blasting, the surface shall be cleaned to remove loose debris and dust before the application of paint.

While too high a blast profile results in an inadequate coating coverage over any high and sharp peaks, leading to premature coating breakdown, an insufficient surface profile may also lead to a redistribution of contaminants over a steel surface, trapping contaminants under the surface as shown in the following sketch.

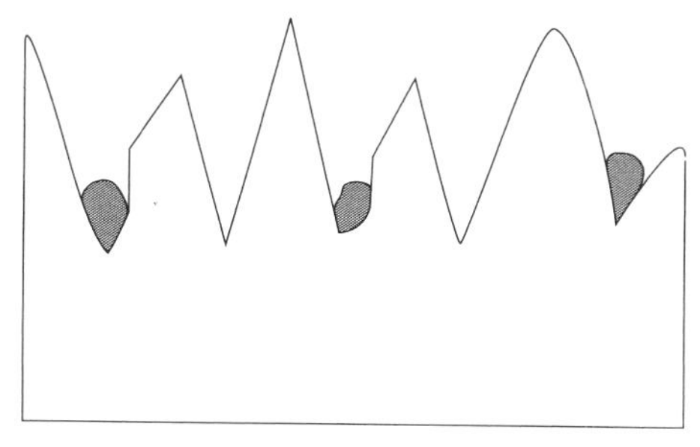

Trapped Contamination

If blasting media is contaminated, the quantity of soluble salts on a steel surface after blasting can be higher than that present before blasting. The quantity of soluble salts in blasting media can be checked by aqueous extraction techniques. Soluble contaminants remaining on a surface can be quantified using commercially available tests.

Advantages and Disadvantages of Abrasive Blasting

Advantages	**Disadvantages**
It can be used to clean large areas of steel.	It is environmentally unfriendly, noisy, and dirty and produces large quantities of dust.
It gives a good profile to steel before painting.	It can leave retained grit in steel, which is not fully over coated. The grit breaks down early in service, giving the appearance and effects of a poor coating system.
It removes rust, old paint, some oils, grease, and soluble salts.	

Blast cleaning is based on the principle of an abrasive jet of particles in a compressed air medium, impinging the metal surface with high velocity, thus removing impurities, mill scales, rust, and old paints. Abrasive blast cleaning is the most thorough and widely used method of surface preparation in the industry. Different degrees of surface cleanliness are possible and depend on the initial condition of the surface and the time it is exposed to an abrasive jet. Apart from cleaning the surface, abrasive particles provide surface roughness to the blasted surface.

However, prior to blasting, the steel surface shall be degreased, and all weld spatters removed. If it is likely that salts, grease, or oil is present on the surface to be blasted, then it might appear that this was removed during blasting, whereas in reality it is not so. Although this is not visible, contamination will remain as a thin layer and will affect the adhesion. Any presence of salts can be checked by measuring the conductivity of water that has been used to wash certain small areas of a blast-cleaned surface.

Furthermore, weld seams, metal slivers, and sharp edges revealed by blast cleaning shall also be ground down to the sound metal layer, as paint coatings tend to run away from sharp edges, resulting in thin coatings and thereby reduced protection at these locations. Weld spatter (not removed after welding till surface preparation) is yet another menace to the coating system, as this cannot be coated evenly. Moreover, weld spatter is loosely adherent and many times causes premature failure of the coating.

Standards for Surface Preparation

Surface appearance resulting from blast cleaning has been defined by several bodies, as detailed in the following table:

United States	ASTM D 2200 and SSPC VIS 1 and 2
Britain	BS 4232
Germany	DIN 18364
Japan	JSRA SPSS 1975
Sweden	SIS 05 5900
ISO	ISO 8501-1
NACE	NACE 1, 2, 3, and 4
Canada	CGSB 31 GP 401 to 404

Of these, the American Standard (SSPC) and ISO are the most extensively used standards in the industry.

Summary of Surface Preparations Techniques and Standards

The following table provides a summary of standards applicable to various cleaning systems mentioned in this chapter.

Sl. No.	Cleaning System	American SSPC SP	NACE	Canadian CGSB	Swedish SIS 05-5900	British	International ISO 8501-1
1	Solvent	SSPC SP 1					
2	Hand tool	SSPC SP 2		31 GP 401	St. 2 (approx.)		St. 2
3	Power tool	SSPC SP 3		31 GP 402	St. 3		St. 3
4	Flame clean (new steel)	SSPC SP 4		31 GP 403			
5	White metal blast	SSPC SP 5	NACE#1	31 GP 404 Type 1	Sa. 3	BS 4232 1st quality	Sa. 3
6	Commercial blast	SSPC SP 6	NACE#3	31 GP 404 Type 2	Sa. 2	BS 4232 3rd quality	Sa. 2
7	Brush off blast	SSPC SP 7	NACE#4	31 GP 404 Type 3	Sa. 1	Light blast to brush off	Sa. 1
8	Pickling	SSPC SP 8					
9	Weather and blast	SSPC SP 9					
10	Near white blast	SSPC SP 10	NACE#2		Sa. 2½	BS 4232 2nd quality	Sa. 2½
11	Power tool to bare metal	SSPC SP 11					

Initial Condition of Steel as per ISO 8501-01

Note: Chart courtesy M/s HMG Paints Ltd, UK.

Sl. No.	Rust Grade	Description	Picture
1	Rust Grade A	Steel covered completely with adherent mill scale and with if any little rust.	
2	Rust Grade B	Steel surface which has begun to rust and from which the mill scale has begun to flake.	
3	Rust Grade C	Steel surface on which the mill scale has rusted away or from which it can be scrapped, but with little pitting visible to naked eye.	
4	Rust Grade D	Steel surface on which the mill scale has rusted away and on which considerable pitting is visible to naked eye.	

Surface Preparations According to ISO 8501-01 Based on Rust Grades

Cleaning Standard	Initial Steel Condition			
	Rust Grade A	Rust Grade B	Rust Grade C	Rust Grade D
St. 2- Hand Tool	Not Applicable			
St. 3- Power Tool	Not Applicable			
Sa. 1- Brush off Blast	Not Applicable			
Sa. 2- Commercial Blast	Not Applicable			

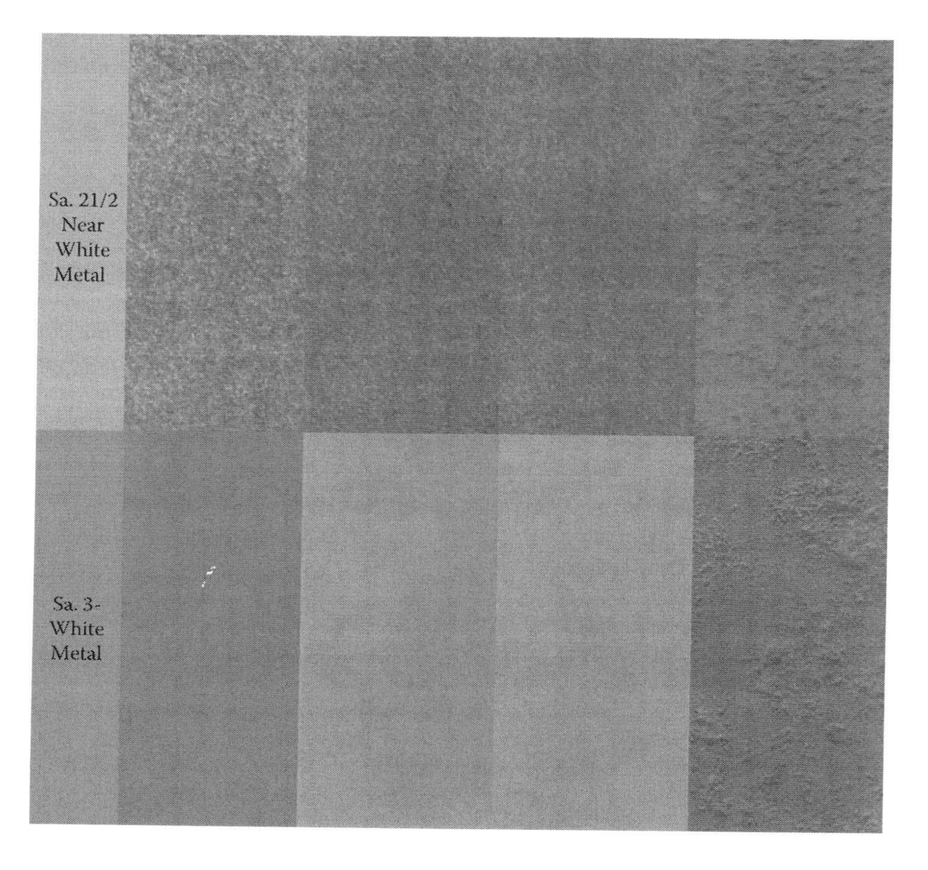

Types of Abrasives

The surface profile obtained during abrasive blasting is important and depends on the abrasive used, air pressure, and technique used in the blasting operation. Too low a profile might not provide a required anchor for coating, whereas too high a profile might result in uneven coverage at high sharp peaks, possibly leading to a premature coating failure, particularly in the case of thin film coatings like blast primers. The following table provides an overview of achievable roughness when different kinds of abrasives are used in blasting.

Sl. No.	Type of Abrasive	Mesh Size (Maximum)	Height of Profile
1	Very fine sand	80	37μ (1.5 mils)
2	Coarse sand	12	70μ (2.8 mils)
3	Iron shots	14	90μ (3.6 mils)
4	Copper slag	1.2–2.0 mm grain size	75–100μ (3–4 mils)
5	Iron grit No. G 16	12	200μ (8.0 mils)

The profile or roughness of a prepared surface can be considered as the most important factor in providing proper anchorage of paint systems. Mineral slag blasting grit generally gives faster rates of cleaning and lower health risk (from shattered grit) than that caused by sand. Grit also gives effective cleaning, especially for pitted substrates. Some grades of grits can be recycled, whereas some cannot be.

Surface Profile

The surface profile indicates the roughness of the blast-cleaned surface. The surface profile is an independent factor and has no connection with standards of cleanliness. Profile roughness obtained during blasting is important and depends on the abrasive media, air pressure used, and technique of blasting. To specify roughness, a variety of values are used such as Rz, Rt, and Ra.

Rz = Average peak to valley height = blasting profile
Rt = Maximum peak to valley height
Ra = Average distance to an imaginary center line that can be drawn between peaks and valleys = CLA = center line average (ISO 3274)
Rz = 4 to 6 times CLA (Ra) (also referred as blasting profile)

13.3.7 Spot Blasting

This is an abrasive, localized preparation process commonly used for outside surfaces during repair and maintenance work, when patches of localized corrosion have occurred. Care shall be taken to avoid the following situations:

- Undercutting and loosening of paint edges around the cleaned spot shall be avoided. Edges shall be feathered wherever and to the extent possible.
- Stray abrasive particles might damage the surrounding sound paint in confined spaces, and such damages if caused shall be remedied and repaired appropriately.
- Blasting shall be discontinued while moving from one spot to the next; blast media should not be trailed over the entire surface. Any damage caused in this way shall be repaired appropriately.

Spot blasting can be used to yield surfaces that are cleaned to Sa. 2 or better, but often surrounding intact areas are prepared with stray grit blasting. These areas also shall be treated as inclusion in grit blasting in the final coating system as well to avoid premature failure of coating in this region. Therefore, it is essential that areas for spot blasting are properly marked and blasted and followed by mechanical feathering of the adjacent area of the existing coating (which is intact), using a rotary disc or sander.

13.3.8 Hydroblasting or Water Jetting

While dry abrasive blasting is the most commonly used method of surface preparation, government and local regulations are continuously changing and require the development of more environmentally sensitive and user-friendly methods of surface preparation. In this context, hydroblasting (also known as hydrojetting, water blasting, and water jetting) is becoming an increasingly viable means to accomplish required surface preparation, the standards for which are in the pipeline.

It shall be noted that hydroblasted surfaces are visually much different from those produced by abrasive cleaning or power tools, and surfaces often appear dull or mottled after initial cleaning is completed.

One drawback of hydroblasting is the formation of flash rust (also called flash back or gingering) after blasting. Heavy rust formed in a short time period is indicative of residual salt on the steel, and reblasting is necessary before painting. However, light rusting is generally acceptable to most of the paint manufacturers for their products.

Hydroblasting does not produce a profile on the steel surface as compared with abrasive blasting. It does however remove rust and loose paint, as well as soluble salts, dirt, and oils, from steel to expose the original abrasive blast surface profile, including the profile produced by corrosion and mechanical damage, whereas ultra-high-pressure hydroblasting can remove adherent paint as well from a steel surface.

The terms *water washing* (usually used to remove salts, slimes, light fouling, etc., found on tanks under use) and *hydroblasting* (used to remove rust and paint) can easily be confused. The following pressure guidelines might be useful in distinguishing these two processes:

Low-pressure water washing and cleaning	Pressures less than 68 kg/cm^2 (1,000 psi).
High-pressure water washing and cleaning	Pressures between 68–680 kg/cm^2 (1,000–10,000 psi).
High-pressure hydroblasting	Pressures between 680–1,700 kg/cm^2 (10,000–25,000 psi).
Ultra-high-pressure hydroblasting	Pressures above 1,700 kg/cm^2 (25,000 psi). Most machines operate in the range 2,000–2,500 kg/cm^2 (30,000–36,000 psi).

Inhibitors can be added (sometimes) to water to help prevent flash rusting prior to applying coating. However, they are often ionic in nature and shall be completely removed by further washing before paint is applied. It is also important to ensure that water being used is sufficiently pure and does not contaminate the surface being cleaned.

Advantages and Disadvantages of Hydrojetting

Advantages	Disadvantages
It removes soluble salts from a steel surface.	No surface profile is produced. It relies on original profile, if present.
Water as a cleaning material is generally inexpensive and available in large quantities. (The Middle East is an exception.)	Flash rusting may be a problem in humid environments.
There is no contamination of the surrounding areas because there are no abrasive particles.	The areas behind the angles require particular attention as they are difficult to clean by hydroblasting.
There is no dust.	

Water jetting or hydroblasting as a surface preparation technique is being used more in shipyards. The major advantage of using water as a medium for an abrasive jet is its low impact on the environment and the health of the operating personnel.

13.3.9 Wet Slurry Blasting

Wet abrasive blasting may be performed with low- or high-pressure freshwater to which a relatively small amount of abrasives are introduced, and in some cases inhibitors are also added to prevent flash rusting. However, as a general rule, it is recommended not to use inhibitors when cleaning areas are to be immersed during service. This reduces airborne dust and sand; however, the surface needs rinsing after blasting to remove sand and other debris.

13.3.10 Sweep Blasting

In sweep blasting, a jet of abrasive is swept across the surface of steel rather than focused on one area for any period of time. In other words, sweep blasting is nothing but treatment of a surface by quickly passing a jet of abrasive across a surface, which is typically used as a tool to get some surface roughness on an existing, firmly adhering coating in order to facilitate intercoat adhesion. The level of effectiveness of this process is highly dependent on the skill of the operator, type of surface, and particle size of abrasive used. In general, a fine grade of abrasive (0.2 mm to 0.5 mm) is recommended, as larger particles sizes tend to destroy the existing coating too much.

Three major types of sweep blast are in common use:

- Light sweeping is used to remove surface contamination or loose coatings. It is also used for etching existing coatings to improve adhesion. Fine abrasive (0.2 mm–0.5 mm) is commonly used for etching.

- Heavy or hard sweeping is used to remove an old coating or rust back to the original shop primer or bare steel.
- Sweeping shop primers in new construction are used to partially remove shop primer to an agreed standard, immediately prior to overcoating.

13.3.11 Surface Preparation for Other Metals

Aluminum

The surface shall be clean and dry. Any corrosion salts present on the metal surface shall be removed by light abrasion and water washing. The cleaned surface shall then be abraded or very lightly abrasive blasted using low pressure and nonmetallic abrasive (e.g., garnet). Alternately, aluminum can be etched by using an acidic solution or etch primer.

Galvanized Steel

The surface shall be dry, clean, and free from oil and grease before galvanizing. Degreasing requires some effort to obtain a clean surface, as zinc corrosion products can trap grease and other contaminants. Any white zinc corrosion products also shall be removed by high-pressure freshwater washing or freshwater washing with scrubbing.

Sweep blasting and abrading are suitable preparation methods, but freshwater washing shall also be used to remove soluble salts. An etch primer can also be used after cleaning to provide a key for further coatings. Many coatings based on nonsaponifiable polymers can be applied directly to a galvanized surface prepared in this way. Paint companies shall be consulted on suitable preparation methods, primers, and coatings for galvanized steels.

When sweep blasting is not feasible, then an acid etch solution or etch primer shall be used to passivate the surface to provide a key for further paint coating. When steel has been treated with a passivating treatment immediately after galvanizing, this must either be allowed to weather off over a period of several months of exterior exposure or be abraded before application of a coating. In general, etch treatments have no effect on fresh materials of this type.

Stainless Steel

Stainless steels do not require any particularly specialized surface pretreatment prior to coating. These surfaces shall be free from oil, grease, dirt, and other foreign materials by chemical cleaning. The development of a surface profile on stainless steels is highly recommended to ensure good coating adhesion. A profile depth of 1.5 mils to 3.0 mils is suggested for most of the coating systems on stainless steels. Abrasive blasting is recommended to achieve this profile.

13.4 Edge and Weld Preparation for New Construction

Experience has shown that edges and welds are generally the first areas to show corrosion and coating breakdown. This is due to a number of interrelated processes including surface preparation, coating application, deflection, shear and buckling stresses on edges and welds, and so on. The quality of surface preparation also plays a major role in determining the service lifetime of coatings.

13.4.1 Weld Preparation

The process of welding generally produces some type of slag on the weld itself, together with spatter and fume. Submerged slags may have to be removed manually.

Weld fume (brown stains) and spatter.

Removal of weld spatter is essential, as this material will cause an irregular surface and will result in poor coverage by paint. Spatter is adherent and must be removed by chipping or other mechanical means.

Weld fume must also be removed, as it is loosely adhered to steel, and depending on the type of weld consumable used, fume may contain water-soluble species. If weld fume is over coated, paint may blister and/or peel off from steel while in service. Blisters can also form where shop primer is damaged due to welding of stiffeners on the other side of the plate (as in floating deck welds). This is often referred to as "burn through," and the following photograph is typical of results of such original defects during service when the burn through is not adequately cleaned or removed prior to coating application.

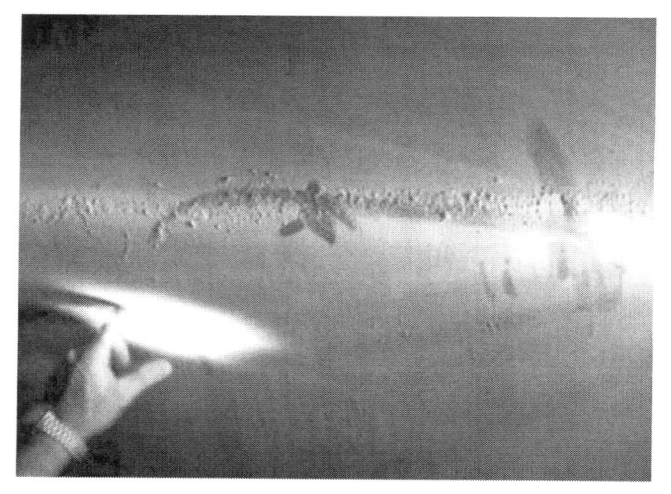

Blistering where a stiffener was welded on the other side of the plate.

The position of welds can also present difficulties to the cleaning and surface preparation process, particularly when the weld forms part of a complex structure such as that in pontoons and floating decks of floating roof storage tanks.

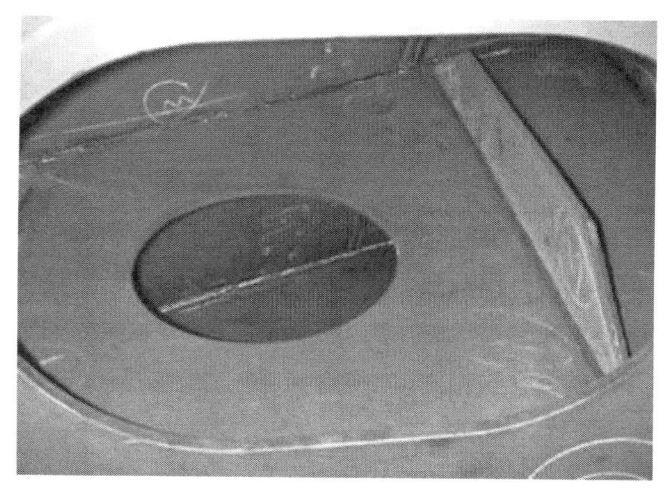

Complex structure, with lower welds only accessible through the hole.

In many cases, only an abrasive or water-blasting process will provide an efficient cleaning of the weld; however, this may not be practical in some new constructions or during maintenance. Welds shall be prepared efficiently so that the possibility of creating voids under coating is eliminated. Porosity can also occur in welds, and this may not become visible until the weld has been blasted cleaned.

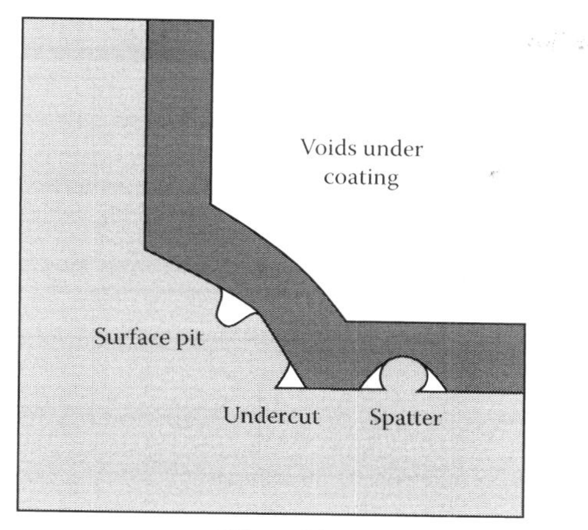

Schematic fillet weld showing typical defects that form voids under the coating.

13.4.2 Edge Preparation

After application while the coating is still in liquid form, there is a tendency for many coatings to pull back from sharp edges, leaving a very thin layer of paint that can quickly break down in service. Grinding profiles into the edges of cutouts, drainage holes, and so on, as shown in the following figure, greatly improves adhesion and coverage of coating around such edges. Rounded edge preparation shall generally provide the most effective service performance from coatings. Two or three passes of a grinding disc over the cut edge shall provide a better preparation from coating point of view. The addition of a stripe coating to the edges is also beneficial in providing long-term protection.

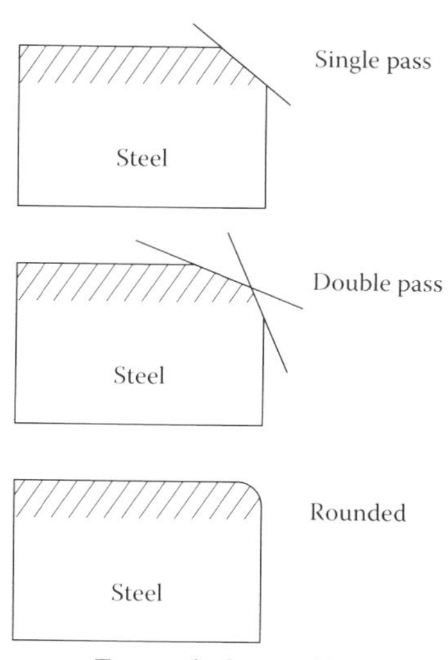

Types of edge profile.

13.5 Paints and Their Uses

While selecting the type of paint, it is necessary to consider the exposure condition of each plant or equipment. The following table provides a few typical conditions and types of paint recommended for those conditions.

Sl. No.	Generic Name of Paints	Paint Film Performance (Resistance to Environment)						Exposed Conditions and Applications					
		Weather	Water	Acid	Alkali	High Temperature	Impact	Nonpolluted and Inland	Polluted and Inland	Nonpolluted and Coastal	Polluted and Coastal	Seawater Immersion	Chemical Immersion
1	Alkyd paints	VG	F	F	F	80°C	G	[a]	—	—	—	—	—
2	Phenolic paints	G	G	VG	F	80°C	G	[a]	—	—	—	—	—
3	Chlorinated rubber paints	VG	VG	G	G	80°C	G	[a]	[a]	[a]	[a]	—	—
4	Vinyl paints	VG	VG	VG	E	80°C	F	[a]	[a]	[a]	[a]	—	—
5	Epoxy paints	VG	VG	VG	VG	60°C	E	[a]	[a]	[a]	[a]	—	—
6	Epoxy paints for water immersion	—	E	E	E	100°C	E	[a]	[a]	—	—	[a]	
7	Epoxy paints for chemical immersion	—	E	E	E	100°C	E	[a]	[a]	—	—	—	[a]
8	Modified epoxy paints	F	E	E	E	100°C	VG	[a]	[a]	[a]	[a]	[a]	[a]
9	Nontar epoxy paints	F	E	E	E	100°C	VG	[a]	[a]	[a]	[a]	—	—
10	Polyurethene paints	E	VG	VG	VG	100°C	E	[a]	[a]	[a]	[a]	—	—
11	Fluoro plymer paints	E	VG	VG	VG	100°C	E	[a]	[a]	[a]	[a]	—	—
12	Silicone paints	VG	VG	G	G	200°C–600°C	G	[a]	[a]	[a]	[a]	—	—
13	Inorganic zinc rich paints	VG	VG	G	G	400°C	VG	[a]	[a]	[a]	[a]	—	—

Note: Charts courtesy M/s Kansai Paints, Japan. E = excellent, G = good, P = poor, F = fair, VG = very good.

[a] Well suited.

13.6 Effective Life of Coating and Film Thickness

A coat of paint inevitably bears some defects and a partial deficiency in film thickness even when the paint is applied thickly. To reduce these defects, it is recommended to apply two or more coats of rust preventive primer before the topcoat. Thicker primer film, if it bears no defects, is more rust preventive. The following table shows the durability of paint film in various weathering conditions. These paint films consist of oil-based rust preventive primers and alkyd topcoats having a total film thickness of about 125 microns.

Weather Conditions and Durability of Paint Film

Sl. No.	Condition	Average Repainting Interval
1	Seaside	3.9 years
2	Industrial	6.0 years
3	Rural	6.9 years
4	Mountain region	7.8 years

Note: Excerpts from Report of Japan National Railway Technical Research Institute No. 892 (Feb. 1974).

In the painting and coating manual of the Royal Dutch Shell Group, coating intervals for maintenance of its paint are shown in the following table. These intervals have been decided for the standard painting system, which consists of two coats of red oxide oil alkyd primer and two coats of aluminum paint. The total thickness of the film is at least 125 microns.

Repainting Interval for Plants

Sl. No.	Condition Structure	Temperate Climate	Tropical or Semitropical Climate	Exposure to Salt-Bearing Winds	Exposure to Frequent Sandstorms and Marine Atmosphere
1	Storage tanks				
2	Walls	10–12 years	8–10 years	4–6 years	2–4 years
3	Roofs	7–10 years	4–6 years	3–5 years	2–4 years
4	Pipes and other structures	8–10 years	6–8 years	3–5 years	2–4 years

Note: Excerpt from Royal Dutch Shell Group manual, Painting and Coating DEP30.48.00 10-Gem. (Jan. 1973).

13.7 Other Requirements by Clients for Surface Preparation and Lining and Painting of Tanks

13.7.1 General

All tank painting and paint testing shall be in accordance with AWWA D102, Steel Structures Painting Council Specification SSPC PA1, approved paint manufacturer specifications, and as specified herein.

Each system of painting shall be from a single manufacturer. Usually a list of acceptable manufacturers is also provided in the contract.

Preconstruction primers may be utilized in the fabrication process to preserve the blast profile and cleanliness. In the field, weld seams and abraded areas shall be cleaned on a spot basis. The remaining sound primer shall be cleaned to remove dirt and other contaminants. After cleaning, a specified coating system shall be applied in its entirety in the field to obtain the required dry film thickness for coating.

No paint shall be applied when the temperature of the surface to be painted is below the minimum temperature specified by the paint manufacturer or less than 5 degrees above the dew point temperature. Paint shall not be applied to wet or damp surfaces or when the relative humidity exceeds 85% unless allowed by the manufacturer's data sheets. Follow the manufacturer's recommendations for the specific paint system used.

After erection and before painting, remove slag, weld metal spatter, and sharp edges by chipping or grinding. All surfaces that have been welded, abraded, or otherwise damaged shall be cleaned and primed in the field in accordance with the paint system requirements.

All areas blasted in the field shall be coated the same day before any rusting occurs.

13.7.2 Lettering and Logo

Lettering and logo design, size, and location shall be as indicated on the drawings issued by clients in this regard. The lettering and logo shall be applied using one coat of aliphatic acrylic polyurethane based on the aliphatic isocyanate curing agent in the required color.

13.7.3 Safety Precautions for Blasting and Painting Equipment

- Blast cleaning pot, nozzles, and spray painting machines shall be earthed (grounded, spark proofed) to prevent buildup of an electrostatic charge.

- All blasting equipment shall be periodically inspected and certified by an authorized, competent third-party inspector. Pressurized hoses shall be tested and certified for maximum safe working pressure. Hoses and other pressure items shall be checked periodically to make sure that any damage or loss of electrical conductivity will not lead to a safety hazard.
- Pressurized and blasting guns shall be compulsorily equipped with an operational "dead-man switch" located immediately behind the gun. No maintenance job shall be performed on a pressurized or energized blasting pot.
- Warning signs shall be provided all around the area where shot and grit blasting is carried out.
- Air masks (hoods) and all other appropriate personal protective equipment (PPE) shall be worn by painting and blasting crew members while blasting or spray painting.
- Operators shall remove their supplied air breathing equipment only when they are far away from the work location, as dust and other contaminants can remain suspended in the air for long periods of time.
- Operators shall wear coveralls and other PPE that provide suitable protection from rebounding abrasives and especially shall wear work gloves that protect the full forearm.
- All personnel involved in blasting and painting operations shall be provided the necessary awareness and training on hazards and risks of their operations and respective risk control measures. All personnel shall be able to produce valid documentary evidence to this effect at the time of evaluations or audits at the site.

13.8 Commonly Used Color Schemes to Reduce Vapor Loss

Tank external	Aluminum	RAL 9006 leafing grade
	White	RAL 9010
Structural steel	Aluminum	BS 4800

13.9 Commonly Used External Coating Systems (for Atmospheric Temperatures)

Sl. No.	Description	Requirement	Remarks
Tank			
1	Surface preparation	SA 2 ½	Profile 35-50 μ
2	Primer	Zinc rich epoxy	70 μ dry film thickness (DFT)
3	Intermediate coat	Two pack hi build micaceous iron oxide (MIO) epoxy	125 μ DFT
4	Finish coat	Two pack acrylic poly urethane (PU) based on aliphatic isocyanate curing agent	50 μ DFT with a total of 245 μ
Ladders, Platforms, Handrails, Stairways, Walkways, etc.			
5	Surface preparation	SA 2 ½	Profile 35-50 μ
6	Coating	Hot dip galvanizing to ISO 1461/ ASTM A 123	Minimum coating weight 610 gm/m^2
Alternative to Galvanizing			
7	Surface preparation	SA 2 ½	Profile 35-50 μ
8	Primer	Zinc rich epoxy	70 μ DFT
9	Intermediate coat	Two pack hi build MIO epoxy	125 μ DFT
10	Finish coat	Two pack acrylic PU based on aliphatic isocyanate curing agent	70 μ DFT with a total of 265 μ

13.10 Commonly Used Internal Lining Systems (for Atmospheric Temperatures)

Sl. No.	Description of Tank or Tank Part	Tank Bottom and Lower Shell	Shell and Structure	Nozzles and Internal Piping	Roof Underside and Roof Structure
1	Brackish and fire water tanks	FGRL up to 600 mm	PE/HBE	PE/HBE	PE
2	Effluent, source, treated water tanks	FGRL up to 600 mm	GF(1)	PE/HBE	PE
3	Dry crude tank	FGRL up to 600 mm	PE/HBE	PE/HBE	PE
4	Wet crude tank	FGRL up to 3,000 mm[2]	PE/HBE	PE/HBE	PE

Sl. No.	Description of Tank or Tank Part	Tank Bottom and Lower Shell	Shell and Structure	Nozzles and Internal Piping	Roof Underside and Roof Structure
					(Continued)
5	Floating roof tank	FGRL up to 600 mm	No coating	HBE (within 600 mm)	Not applicable
6	Potable water tank	SFE	SFE	SFE	SFE

Note: FGRL = fiberglass reinforced lining, PE = phenolic epoxy, HBE = high build epoxy, GF = glass flake, SFE = solvent-free epoxy (suitable for drinking water).

[1] If GF cannot be applied on a structure due to its shape, then PE/ HBE can be applied.

[2] If the water level is higher than 3.0 m, then apply FGRL up to 0.60 m with the remaining shell height lined with GF.

[3] All tank bottom plate soil sides shall be coated with HBE.

13.11 Inspection and Tests with Recommended Frequency for Surface Preparation and Lining and Painting

Sl. No.	Type of Test	Method	Frequency	Acceptance Criteria	Consequence
1	Environmental conditions	Ambient temperature Steel temperature Relative humidity Dew point	Before start of each shift Minimum twice per shift	Specification	No blasting or coating
2	Abrasive material	Certification	Spot check	Specification	No blasting
3	Visual examination (blasting)	Sharp edges Spatter and slivers Rust grade, etc. As per ISO 8501-3	100% of all surfaces to be coated	No defects (refer to specification for details)	Repair
4	Cleanliness	Visual as per ISO 8501-1 Dust as per ISO 8502-3	100% surface Spot check	Specification Specification	Reblasting Recleaning and retesting
5	Salt test	ISO 8502-6 ISO 8502-9	Spot check	Maximum conductivity corresponding to 20 mg/ m^2NaCl	Repeated washing with potable water
6	Roughness	Comparator, stylus instrument, or Testex tape (ISO 8503)	Each component or once per 10 m^2	Specification	Reblasting
7	Curing test (for zinc silicate)	ASTM D 4752	Each component or once per 100 m^2	Rating 4 to 5	Allow to cure

Sl. No.	Type of Test	Method	Frequency	Acceptance Criteria	Consequence
8	Visual of coating	To determine curing Contamination Solvent retention Pinholes and popping Sagging and surface defects	100% surface after each coat	Specification	Repair defects
9	Holiday detection	ISO 29601 voltage	100% surface	No holidays	Repair and retest
10	Film thickness	ISO 19840 calibration on smooth surface	ISO 19840	ISO 19840 and coating system data sheet	Repair, apply additional coats, or recoat as appropriate
11	Adhesion	ISO 4624 using equipment with an automatic centered pulling force and carried out after coating is fully cured	Spot check	Based on painting scheme	Coating rejected

14

Documentation

14.1 General

This chapter describes the various documents to be included in the manufacturer's record book (MRB) for storage tank construction, which shall be the authentic document pertaining to the tank and contain all salient details about the tank, right from purchase or work order to design, materials, construction, inspections and tests, surface preparation and painting, and so on for future reference during the service life of the tank. As this document does not have any immediate use, often its due importance is not recognized well. But when some modification or rectification is needed during service, many technical details will be required to carry out the design for proposed modification, alteration, or repair. If due attention was not paid while finalizing the contents of the MRB, salient information may not be available to help carry out the design for the proposed modification. Therefore, it is equally important to get a compilation of all relevant details and activities (especially inspections and tests) that were part of the construction process from the contractor. For the contractor to compile it properly and systematically, it is important that the requirements are provided to the contractor, preferably as part of the contract, so that the contractor can gear up the systems to generate such a document to meet the expectations of the client. In the following sections, a comprehensive list of documents to be included in the MRB is provided. Depending on the specific requirements and applications, additional documentation might be required. However, for all general purposes, the documents discussed shall suffice.

14.2 Manufacturer's Record Book

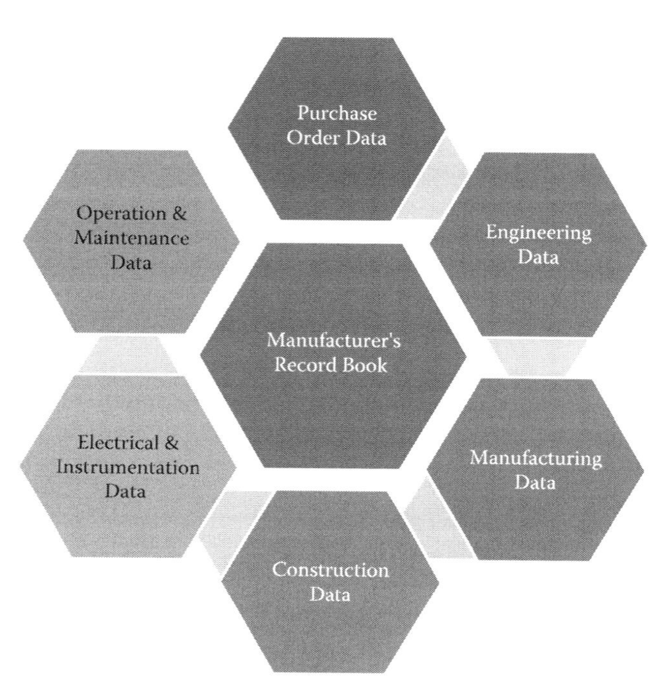

The MRB can be structured into six sections. The appended table indicates a typical list of contents generally required in the oil and gas industry under each of the six sections mentioned. This may be modified as required by clients according to their requirements. The list is prepared with a view to include floating roof tanks as well under its umbrella, which would not be applicable for normal fixed cone roof tanks. Though it is true that many of the documents listed may not be of any use in the entire service life of the tank, still it is recommended to make it comprehensive enough to enable the client to proceed with modifications during service without seeking help from the previous contractor who originally constructed the storage tank. Many times, the original contractor might be stationed elsewhere, and getting their help and assistance would be costly on account of time and money. If all salient information of the existing tank is available with the client, the client can get the modifications done through local contractors, with in-house engineering support provided by the client. Therefore, it is always better to collect maximum information with regard to design, procurement, construction, inspection and testing, surface preparation and painting, and so on from the contractor as part of the contract.

By incorporating basic requirements for the preparation of the MRB in the contract document, from day one, the contractor is aware of the types of documents to be included in the MRB and is in a position to take proactive actions to get such documents from subvendors as well in time. Furthermore, the contractor shall start compiling the MRB in pace with the physical progress of works at the site so that approval and handing over of the MRB to the client is possible soon after commissioning the tank.

The tank construction package may even consist of other related works within the tank farm, such as piping, electrical and instrumentation works, a dyke wall, pipe supports, fencing, a surveillance system, a public address system, and so on. However, the following table includes only documents pertaining to the tank proper that are covered by the jurisdiction of API 650.

14.3 Suggested Contents for MRB

Sl. No.	Description of Document	Document Status		MRB	Size
		Original	As-Built		
1.0	*Purchase Order Data*				
1.1	Purchase Order	X		X	A4
1.2	Technical Procurement Specification	X		X	A4
1.3	Data Sheet (Mechanical and Others)	X		X	A4
1.4[a]	Scope Drawing	X		X	A4[a]
1.5	Change Order (If Any)	X		X	A4
1.6	Deviation or Concessions (If Any)	X		X	A4
1.7	Minutes of Meetings (Kick Off and Preinspection)	X		X	A4
2.0	*Engineering Data*				
2.1	Detailed Data Sheet		X	X	A4
2.2	Design Calculations		X	X	A4
2.3[b]	Drawings				
2.4[b]	General Arrangement Drawing		X	X	A3
2.5[b]	Bottom Plate Layout and Weld Details		X	X	A3
2.6[b]	Bottom Plate Cutting Plan		X	X	A3
2.7[b]	Shell Development and Shell Weld Details		X	X	A3
2.8[b]	Shell Plate Cutting Plan		X	X	A3
2.9[b]	Roof Plate Layout and Weld Details		X	X	A3
2.10[b]	Roof Plate Cutting Plan		X	X	A3
2.11[b]	Dome Roof Details		X	X	A3

(Continued)

Sl. No.	Description of Document	Document Status		MRB	Size
		Original	As-Built		
2.12[b]	Roof Structure Details		X	X	A3
2.13[b]	Draw Off Sump Details		X	X	A3
2.14[b]	Nozzle and Manway Schedule		X	X	A3
2.15[b]	Special Notes		X	X	A3
2.16[b]	Nozzle Weld Details (Shell and Roof)		X	X	A3
2.17[b]	Clean-out Door Details		X	X	A3
2.18[b]	Manway Weld Details (Shell and Roof)		X	X	A3
2.19[b]	Manway Davit Details		X	X	A3
2.20[b]	Wind Girder Details		X	X	A3
2.21[b]	Stairway Details		X	X	A3
2.22[b]	Ladder and Platform Details		X	X	A3
2.23[b]	Handrail Details		X	X	A3
2.24[b]	Floating Roof Details		X	X	A3[d]
2.25[b]	Pontoon Fabrication Details		X	X	A3[d]
2.26[b]	Pontoon Support Column Details and Locking System		X	X	A3[d]
2.27[b]	Details Antirotation Device		X	X	A3[d]
2.28[b]	Roof Drain Details		X	X	A3[d]
2.29[b]	Details of Seal Arrangement		X	X	A3[d]
2.30[b]	Details of Inlet Distributor Pipes		X	X	A3
2.31[b]	Details of Suction Header		X	X	A3
2.32[b]	Details of Other Internal Attachments		X	X	A3
2.33[b]	Details of Other External Attachments		X	X	A3
2.34[b]	Details of Piping Clips		X	X	A3
2.35[b]	Details of Deflectors		X	X	A3
2.36[b]	Details of Foam System		X	X	A3
2.37[b]	Anchor Chair Details		X	X	A3
2.38	Name Plate		X	X	A4
2.39	Instrument Data Sheets		X	X	A4
2.40	Instrument Drawings		X	X	A4
2.41	Electrical Drawings		X	X	A3
2.42	Cathodic Protection Details		X	X	A3
3.0	*Manufacturing Data*				
3.1[a]	Construction Quality Plan		X	X	A4[a]
3.2	WPS/PQR/WQT Records		X	X	A4
3.3	Tank Erection Procedure		X	X	A4
3.4	Welding Consumables Control Procedure		X	X	A4
3.5	Hydrostatic Test Procedure for Completed Tank		X	X	A4
3.6[c]	Roof Floatation Test Procedure		X	X	A4[d]
3.7[c]	Water Accumulation Test Procedure on Floating Roof		X	X	A4[d]

Sl. No.	Description of Document	Document Status		MRB	Size
		Original	As-Built		
3.8[c]	Pontoon Puncture Test Procedure for Floating Roof		X	X	A4[d]
3.9[c]	Drain Pipe Test Procedure		X	X	A4[d]
3.10	Water Draining, Cleaning, Drying, Disposal Procedure		X	X	A4
3.11	Pneumatic Test Procedure for Roof Welds		X	X	A4
3.12	Pneumatic Test Procedure for Reinforcement Pads		X	X	A4
3.13	Procedure for Calibration of Welding Equipment		X	X	A4
3.14	Procedure for Calibration of Electrode Oven		X	X	A4
3.15	Hardness Test Procedure		X	X	A4
3.16	PWHT Procedure		X	X	A4
3.17	Vacuum Box Test Procedure		X	X	A4
3.18	Settlement Measurement Procedure		X	X	A4
3.19[a]	Inspection and Test Plan (Mechanical Works)		X	X	A4[a]
3.20	Surface Preparation and Coating Procedure for Internal		X	X	A4
3.21	Surface Preparation and Coating Procedure for External		X	X	A4
3.22	Inspection and Test Plan (Surface Preparation and Painting/Coating)		X	X	A4
3.23	Liquid Penetrant Test Procedure	X	X	X	A4
3.24	Magnetic Particle Test Procedure	X	X	X	A4
3.25	Ultrasonic Test Procedure	X	X	X	A4
3.26	Radiographic Test Procedure	X	X	X	A4
3.27	Visual Examination Procedure	X	X	X	A4
4.0	*Construction Data*				*A4*
4.1	Material Summary Report and Material Map			X	A4
4.2	For Bottom Plate			X	A4
4.3	For Shell			X	A4
4.4	For Roof			X	A4
4.5	Material Test Certificate (Pressure Parts Like Nozzles Manways and Internals against Each of the Attachments)			X	A4
4.6	Bottom Plate (BP) Weld Map			X	A4
4.7	Welding Inspection Summary and Reports (BP)			X	A4
4.8	NDT Summary and Reports (BP)			X	A4
4.9	Vacuum Box Test Summary and Reports (BP)			X	A4
4.10	Shell Weld Map			X	A4
4.11	Welding Inspection Summary and Reports (Shell)			X	A4

(Continued)

Sl. No.	Description of Document	Document Status		MRB	Size
		Original	As-Built		
4.12	NDT Summary and Reports (Shell)			X	A4
4.13	Vacuum Box Test Summary and Reports (BP)			X	A4
4.14	Manufacture of Subassemblies Like Nozzles, Manways, and Other Appurtenances			X	A4
4.15	Welding Inspection Summary and Reports against Each Item			X	A4
4.16	NDT Summary and Reports			X	A4
4.17	Clean-out Door Welding Inspection Report			X	A4
4.18	NDT of Clean-out Door			X	A4
4.19	PWHT Report for Clean-out Doors			X	A4
4.20	Welding Inspection Reports for Clean-out Door to Shell			X	A4
4.21	NDT Reports of Clean-out Door to Shell			X	A4
4.22	Shell Dimensional Report (Each Shell Course)			X	A4
4.23	Nozzle/Manway Attachment			X	A4
4.24	Welding Inspection Report			X	A4
4.25	NDT Report			X	A4
4.26	Pad Air Test Reports as Applicable			X	A4
4.27	Pressure Gauge Calibration Reports			X	A4
4.28	BOM Check and Clearance for Hydrostatic Test Report			X	A4
4.29	Quality of Water (Test Report)			X	A4
4.30	Water Fill Report			X	A4
4.31	Hydrostatic Test Report			X	A4
4.32	Settlement Report and Cosine Curve			X	A4
4.33	Pneumatic Test Report (for Roof)			X	A4
4.34	Draining Off Report			X	A4
4.35	Cleaning and Drying Report			X	A4
4.36	Signed Off ITP for Erection Work			X	A4
4.37	Surface Preparation and Painting/Lining Reports			X	A4
4.38	Adhesion/Other Coating Inspection Test Reports			X	A4
4.39	Signed Off ITP for Surface Preparation, Painting/Lining			X	A4
4.40	Tank Calibration Reports			X	A4
4.41	Boxing Up Report			X	A4
4.42	Inertization Report			X	A4
4.43	Handing Over Report			X	A4
5.0	*Electrical and Instrumentation Data*				
5.1	Installation and Test Reports for Instruments			X	A4
5.2	Calibration Report for Instruments			X	A4
5.3	Cathodic Protection Completion Report			X	A4

Sl. No.	Description of Document	Document Status		MRB	Size
		Original	As-Built		
5.4	Cathodic Protection Test Reports			X	A4
5.5	Electrical Installation Commissioning Report			X	A4
6.0	*Operation and Maintenance Data*				
6.1[a]	Periodic Inspection and Maintenance Requirements			X	A4[a]
6.2	Operating Instructions			X	A4
6.3	Spare Parts List			X	A4

Note: 1. For documents under construction data, the size may be A4 landscape or portrait as required.
2. Minimum documents were provided only under Sections 5 and 6 for the sake of completeness.
3. Apart from one hard copy, the MRB shall be made available as a soft copy as well in PDF format.
4. The documents shall be searchable in PDF.
5. Classification of documents may be interchanged according to the system of the client.

[a] A4 landscape.
[b] Items listed shall be provided as autocad (ACAD) files in addition to the PDF for the MRB.
[c] May be included in one procedure applicable to floating roof tanks.
[d] Applicable only to floating roof tanks.

14.4 Documents Not Specified by Standards

Some of the documents indicated in the previous table are not considered relevant by code. However, these documents are specified in the table to corroborate each and every document included in the MRB to the respective component shown in the design and drawing, including its traceability.

For example, under purchase order data, as Item 1.3, the original data sheet given to the contractor along with the purchase order is specified as a content of the MRB. Subsequently under engineering data, as Item 2.1, the as-built data sheet prepared by the contractor is also specified. Though this requirement looks like a repetition, these two documents are included in the MRB for a specific reason. The data sheet prepared by the client or its consultant provides all absolutely essential requirements from a service point of view, whereas the data sheet submitted by the contractor contains more details based on design considerations, constructability, code requirements, and other constraints such as availability, cost, and so on. Because of these diversified outlooks, many more salient details could be seen in the contractor's data sheet compared to those in the client's data sheet provided with the purchase order. To understand this evolution that occurred during detailed engineering, it is essential to have both these documents in the MRB, failing which this history would go unnoticed, which might be useful at a later date for a critical assessment.

Similarly, some more documents such as weld maps, weld and nondestructive testing (NDT) summaries, and so on are also of paramount importance in ensuring smooth corroboration of documents pertaining to each and every part that constitutes a component of the storage tank and ensures completeness of coverage (especially that of NDT and other tests) as envisaged in the contract. While describing the relevance of each document would be impractical, some are explained next since they are considered as key in correlating documents to applicable drawings and to part numbers indicated therein.

14.4.1 Material Summary

The material summary is nothing but a tabulation of a list of part numbers (as per the drawing) against material test certificates (MTCs). In other words, it is a summary statement showing all part numbers connected with a material certificate. With the help of this document, at one glance, the actual material used during construction for a particular part number in the drawing can be traced back. Therefore, the material summary shall be read in conjunction with the MTCs for all pressure parts, so that the MTC of a particular part can be located easily. At a minimum, material certificates pertaining to all pressure containing parts shall be compiled systematically with the help of a summary as mentioned. In addition, material certificates for all internals also shall be compiled in a similar way. Though it is desirable to compile MTCs of structural steel materials as well, depending on the complexity of the structure, this may be included or excluded from the MRB as circumstances warrant and in consultation with the client or consultant.

14.4.2 Weld Map

The weld map is another key document that usually forms a part of many of the reports such as for welding, NDT, vacuum box testing, and visual inspection. In tank construction three separate weld maps are required, one each for the bottom plate, shell, and roof. Other welds can be monitored through weld joint summary reports. The weld map is a pictorial development (representation) of the bottom plate, shell, and roof plate identifying each weld joint with a unique number that has to be used in all subsequent activities and reports.

14.4.3 Weld Summary

A weld summary table is also accompanied by a weld map (for bottom, shell, and roof plates) that depicts welding processes, welding procedure specifications (WPS), and welders deployed for each of the pressure containing weld joints in the tank, including those for nozzles, manways, cleanout doors, and so on. In addition, nonpressure containing welds in internal piping, such as distribution or spreader pipes, are also considered important and included in the weld summary as per practice. While an overall summary of structural welds is required (to be contained in one or two pages), a detailed one as provided for pressure containing parts is not usually considered essential, as this could make the MRB more bulky. However, it shall be the prerogative of the client to ask for such details as well.

14.4.4 NDT Summary with Weld Map

The NDT summary table is also accompanied by a weld map (for bottom and shell plates) that depicts the type of NDT carried out on each joint (with reference to the concerned report number for easy verification) and approximately indicates the location of the NDT as well. The NDT summary shall cover all pressure containing welds and nonpressure containing welds, such as those in internal piping such as distribution or spreader pipes.

14.4.5 NDT Reports

NDT reports shall accompany the NDT summary and weld map. As far as possible NDT reports shall be compiled in the order of report numbers, allowing easy retrieval of the desired report by referring to the NDT summary.

14.4.6 Pad Air Test Reports

All reinforcement pads shall undergo a pad air test. Since many reinforcement pads can be tested in a day, and if the test pressure is the same, it is possible that all such pad tests can be covered by a single report. Usually, one report for shell nozzles and manways and yet another for roof nozzles and manways would suffice, but all this depends on whether the tests are completed on the same day or not.

14.4.7 Tests Summary

It is desirable to provide a summary of inspections and tests carried out on each tank for quick reference in the future.

14.4.8 Dimensional Inspection Reports

The code requires certain dimensions to be maintained within acceptable limits. These acceptance criteria are arrived at based on the philosophy and considerations adopted during formulation of these codes and other practical considerations. However, in practice, service requirements and environmental conditions may call for further restrictions in tolerances specified in the code.

To meet all code and specification requirements, dimensional checking and reporting at the following stages of fabrication is proposed. Stage-wise measurement and recording provides a forewarning with respect to tighter tolerances (if required) to be enforced in subsequent activities to control the dimensions of the completed tank well within the tolerance specified, especially in the case of floating roof tanks.

A few dimensional measurements in the following table are specified in view of periodic inspections that are to be carried out during the service life of the tank, wherein a comparison with the original readings of the same would be of immense value in assessing the remaining life or fit-for-service assessment of the tank.

Sl. No.	Description of Dimension	Location at Which to Be Taken	Frequency	Remarks
1	Bottom plate, outside circumference	On surface of bottom plate	(Twice) After layout and after welding of all lap and butt welds	Through measurement of OD at 4 coordinates and calculation
2[a]	Buckling of bottom plate	After completion of all lap, butt, and fillet welds	Once from inside of tank	Positively after completion of shell to annular fillet weld
3	Shell outside circumference	At both edges of shell at 50 mm from horizontal weld seam and at middle	Each shell course (at setup and after full welding of vertical seams)	Measurement at middle of shell may be dropped if shell course width is less than 3,000 mm
4	Shell vertical seam peaking in or out	At both edges of shell at 50 mm from horizontal weld seam and at middle	At each vertical seam at fit up for unacceptable plate forming and after welding for peaking in or out	Using a sweep board of chord length 2.5 m to 3 m with weld joint at middle of sweep board
5	Banding at shell horizontal joints	On either side of horizontal weld seam at 25 mm from edge of weld	For each horizontal seam after completion of respective horizontal joints	Two measurements to be tabulated against each shell course
6	Profile at curb angle or compression ring	At 50 mm below compression ring or 25 mm below curb angle	After fit up and after welding	
7	Profile at wind girders	At 25 mm below or 50 mm above wind girders	For each wind girder after fit up and after welding	Preferred reading is below wind girder
8	Nozzles (shell)	Orientation Elevation Stand off	For each nozzle at fit up and after welding at four directions	Stand off from shell needs to be recorded; Tilt of nozzle in any plane shall be in accordance with the norms in API 650
9	Manways (shell)	Orientation Elevation Stand off	For each manway at fit up and after welding at four directions	

Sl. No.	Description of Dimension	Location at Which to Be Taken	Frequency	Remarks
10	Clean-out door	Orientation Elevation	For each	
11	Other attachments (shell)	Orientation Elevation	For each item	
12[a]	Roof plate buckling	On surface of roof plate	After layout and after welding of all lap and butt welds	
13	Nozzles (roof)	Orientation Stand off	For each nozzle at fit up and after welding at four directions	Stand off from roof needs to be recorded; Tilt of nozzle in any plane shall be in accordance with the norms in API 650
14	Manways (roof)	Orientation Stand off	For each manway at fit up and after welding at four directions	
15	Other attachments (roof)	Orientation	For each item	
16[a]	Plumbness or verticality of tank	At each shell course and middle at preselected angles and 50 mm below curb angle, wind girder, etc.		
17[a]	Maximum and minimum of annular space of floating deck	Before fill and after test	Through diametrical line at 10 m on periphery with a minimum of four Measurements to be taken at 3 m intervals including initial and final height of pontoon	For floating roof tanks Recommend recording at various levels of flotation, namely, ¼, ½, ¾ and fully attainable

[a] Serves as a reference to measurements made during periodic maintenance.

14.4.9 Prehydrostatic Inspection Report

This report is a verification document for completion of all previous activities. Generally all welding works and NDT shall be completed prior to the hydrostatic test. The purpose of this document is to verify that all such activities are completed by then. If any work is pending, this shall be enclosed as a punch list, which can be done after the hydrostatic testing.

14.5 General Requirement for Contents of MRB

1. Documents forming part of the MRB that require client approval shall have adequate evidence to prove so, especially in the case of engineering documents, which require prior approval of the client or consultant.
2. The document shall be substantially complete before equipment or materials are released from a vendor's premises. However, in the better interest of all, all critical certifications are completed prior to the dispatch of materials from a vendor's works, otherwise rework if required based on the outcome of review comments by the client may eventually turn out to be costly and time-consuming when carried out at the site.
3. The contractor or vendor shall perform checks by its engineering and QA teams prior to submitting documents for client certification. Corrections are not usually permitted during this MRB review; however, inclusion of additional information if considered important may be endorsed on documents.
4. When modifications or revisions arise after submission of the MRB, the contractor or vendor shall update the document (as a new revision) as required in the document control procedure.
5. One hard copy of the document shall be submitted in loose-leaf binders, with proper indexing available in each volume for cross-reference. The recommended paper sizes for the hard copy are specified in the previous table.
6. The hard copy shall be accompanied by a soft copy in PDF format, which shall be converted electronically into PDF format to maintain requisite quality. The soft copy shall be in searchable PDF format. In addition, all drawings shall be submitted in ACAD and data sheets in EXCEL format as well for implementing additions or deletions in the future during the service life of the tank.
7. The MRB shall be compiled on an ongoing basis during manufacture.
8. Duplication of documents within a dossier shall be avoided to the extent possible.
9. All pages within the MRB shall be clearly marked with the contractor job number, item name and tag number, page number, volume number, and so on as applicable.
10. All documents shall be prepared in the English language. In case some of the certifications are issued in languages other than English, an English translation of the same shall be enclosed with original certification with due endorsement of the contractor as a true translation.

11. The MRB shall be compiled according to a proposed MRB index duly approved by the client right at the beginning. After compilation, a single hard copy of the same shall be submitted for review of the client to ensure proper compilation and completeness of the document. The document thus approved by the client shall be submitted as the MRB in the required number of hard and soft copies.

14.6 Records and Reports of Inspections, Tests, and Calibrations

1. When specifying documents to be included in the MRB, attention shall be paid to the distinction between certificates, reports, and records. Reports and records typically include useful data and definitive statements, whereas certificates can consist of unsupported statements certifying that tests, inspections, and calibrations have been carried out with satisfactory results. Such statements from vendors or subcontractors have no value to the client, especially after expiry of the guarantee.
2. Records and reports shall explicitly indicate the concerned item, material, or equipment through the proper identification system such as the item number, serial number, tag number, and so on as required.
3. When inspections or tests are performed on a sampling basis, this shall be made clear in the inspection certificate issued for the lot represented by the sample.
4. When inspections or tests are performed against a written procedure, reference to this procedure shall be present in the test or inspection report.
5. When results of inspections or tests are compared and accepted based on standards or specifications, reference of this document also shall be made on the inspection or test report.
6. Inspection and test reports shall contain details such as performance date and report date, along with a description of the type of test or inspection carried out.
7. When reports consist of more than one page, all pages shall have the report number on them, along with "page x of y" at appropriate places to ensure completeness of the report.
8. Similarly, any attachments to the report also shall be clearly indicated in the report, including information as to the number of pages of each attachment.
9. Any corrections, alterations, or additions made to the report after endorsement by client inspectors shall be clearly traceable and dated.

14.7 Certification for Materials

1. Material traceability is of paramount importance, and as far as possible original certifications shall be maintained in the MRB.
2. When the original MTC is issued for a large quantity (of which only a small portion is required for the tank, such as pipes, fittings, flanges, etc.), the original certificates shall be available with the supplier or trader. Such availability of the original certification shall be ensured through endorsements made by the inspection engineers of the contractor, consultant, or client at the manufacturer's or trader's premises in the respective inspection reports for materials inspected and released from stockists.
3. Materials without any of these certifications shall not be entertained even for very small quantities.
4. It is mandatory that certificates for wrought butt weld fittings be supported by certification for the parent plate or pipe as applicable. However, it is not essential that such supporting certification be original, as long as product specification requirements are established through sample testing after forming the component.
5. Corrections to certificates shall not be acceptable under any circumstances. When errors are found in a certificate, it shall be reissued at the source or else the material shall be rejected.
6. When supplementary tests (tests not carried out by the manufacturer) are carried out on materials with original certification, this shall be reported on separate sheets with cross-reference to the original certification and endorsed by contractor or client inspectors as applicable.
7. Transcribed data on material certificates shall normally be acceptable under the following circumstances:
    a. heat analyses for wrought materials, and
    b. certificates issued by stockists for bolting materials or screwed and socket weld fittings, which contain data and test results taken from manufacturer's certification and certified as having been accurately transcribed.
8. When material is required to satisfy a carbon equivalent limit as determined by the long formula (C + Mn16 + Cr+Mo+V/5 + Ni+Cu/15), then all component elements of this formula shall be determined and reported. It shall not be acceptable to assume zero content for any unreported elements.

9. For nonferrous alloys, where the applicable standard gives a minimum value for the predominant element, it is not acceptable for this element to be certified as "remainder," as this does not take into account the levels of impurities that may be present.

14.7.1 Components Requiring "Material Certification"

1. All components comprising the pressure envelope
2. All nozzle compensation pads
3. Lifting and tailing attachments
4. Components attached to the shell by welding
5. Weld consumables including consumables for overlay cladding
6. All materials requiring impact testing

14.7.2 Contents of "Certification Dossier" for Bought Out Items (as Applicable)

1. Index of the dossier
2. Code data form or certificate of conformity
3. Third-party final report or certificate of inspection
4. Inspection and test plans (ITPs) signed by all inspection authorities
5. List of materials or layout sketch showing the position of the component, cross-referenced to the page numbers of the material certificates and any supplementary reports
6. MTCs including hardness testing as required
7. Material repair records
8. Weld key map and seam identification sketch (When the manufacturer opts to maintain a written record of the work performed by each coded welder in lieu of stamping welders identification against their welds, this record shall be included in the certification dossier. This record need not extend beyond "code" welds.)
9. Welding procedures and qualification records
10. Production test records
11. NDE reports and records (When repairs have been carried out, reports of original examinations shall be included.)
12. Reports of any required special tests (SSC, HIC, etc.)
13. Heat treatment charts and records as applicable
14. Pressure and leak test reports
15. Coating or lining application records and examination reports

16. Nameplate rubbing (facsimile)
17. As-built general arrangement drawing and isometric drawings for piping
18. Records of critical dimensions
19. All design calculations (including relief valves supplied with packaged vessels)
20. Function test records for actuated valves
21. Valve pressure and temperature ratings
22. Special certificates (fire safe, etc.)
23. Bend manufacturing procedures and qualification tests
24. Pump performance test records, vibration records, and function test records for instruments and controls
25. Pipeline welds traceability, pressure testing charts, and gauging, cleaning, and drying records
26. Dimensional records

14.8 Heat Treatment Records and Charts

1. When a heat treatment of materials or components is performed to satisfy code or specification requirements, it shall be sufficient for the manufacturer to declare that the requisite heat treatment was carried out, as in the case of normalizing, wherein temperatures need not be declared. However, for other heat treatments such as tempering and simulated post weld heat treatment (PWHT), heating and cooling rates, soaking temperature, and time need to be specified in the accompanying certificate.
2. When a heat treatment is carried out on materials to meet NACE MR0175/ISO 15156 requirements, sufficient details shall be reported to verify conformance. Hardness testing on heat treated components also shall be required subsequent to the heat treatment, and results of the same shall be in accordance with NACE MR 0175/ISO 15156.
3. When heat treatments are carried out on fabricated components such as clean-out doors, a heat treatment certificate and a chart of a multipoint temperature recorder shall be provided as record, with due endorsement from the contractor and client or consultant.
4. Records and reports shall always have direct reference to heat treated components like part number, equipment tag number, and so on.

5. Actual chart speed, time of start, and heating, soaking, and cooling zones shall be clearly identified on a chart along with the direction of time progression.
6. Location and identification of thermocouples shall be indicated in a sketch provided with the report.
7. When a heat treatment is performed against a written procedure, reference to this procedure also needs to be reflected in the report.
8. As in the case of other inspection reports, both the performance date and the report date shall be clearly indicated in the heat treatment record as well.
9. When records consist of more than one page, all pages shall be numbered as "page x of y" and have the report number on all pages to ensure inclusion of the complete report.
10. When records are allowed in lieu of recorder charts, the actual holding temperature and time shall be reported. For post-weld or stress-relieving treatments, the maximum actual heating and cooling rates shall also be recorded.

14.9 NDT Reports

1. For all NDTs performed at the site, original reports shall be provided with due endorsement from the contractor or consultant or client as applicable.
2. Reports shall contain details of components or welds radiographed with unique identifications provided to each of the weld seams or components.
3. The personnel deputed to carry out NDTs and prepare the report shall be qualified adequately for the envisaged works and provide evidence of proper qualifications.
4. Details of the examination techniques, equipment, consumables, and extent of testing shall be presented in detail in the report. When technique sheets have been separately approved, reports need only contain a reference to these sheets.
5. As in other reports, both performance date and report date shall be clearly indicated.
6. When records consist of more than one page, all pages shall be numbered as "page x of y" and have the report number on all pages to ensure inclusion of the complete report.

7. The report shall clearly indicate the stage at which the NDT was carried out. For example, when a heat treatment or forming is required in manufacturing, NDT reports shall indicate whether the required examination was carried out before or after the heat treatment operation.
8. Acceptance or rejection criteria adopted in the evaluation also shall be indicated in the report.
9. The minimum achieved sensitivity levels or calibration standards along with the acceptance criteria shall be recorded in the report.

15

Formats

15.1 General

For carrying out inspection and tests and recording the findings at each stage of storage tank construction, various formats need to be used. This chapter shows typical formats proposed for storage tank construction, in accordance with code and client requirements. The purpose of these formats is to provide uniformity in the structure, usefulness, and completeness of the information required so these reports are helpful to assess situations that may arise during operation of the storage tank.

15.2 Formats for Fixed Cone Roof Tanks

Serial No.	Description	Document Code
1	Request for Inspection	RFI
2	Foundation Level Inspection Report	FLR
3	Material Inspection Report	MIR
4	Welding Consumable Inspection Report	CIR
5	Welding Consumable Control Log	CCL
6	Material Traceability Report or Material Summary	MTR
7	Shell Plate Bending Inspection Report	PBR
8	Other Components Bending/Forming Inspection Report	OCR
9	Fit Up Inspection Report	FIR
10	Weld Visual Inspection Report	VIR
11	Radiographic Test Report	RTR
12	Ultrasonic Test Report	UTR
13	Magnetic Particle Test Report	MPR
14	Dye Penetrant Test Report	DPR
15	Post Weld Heat Treatment Report	PWH
16	Hardness Survey Report	HSR

(Continued)

Serial No.	Description	Document Code
17	Weld Summary Report	WSR
18	Weld Peaking Control Report	PCR
19	Banding and Circumference Report	BCR
20	Plumbness Inspection Report	PIR
21	Roundness Inspection Report	RIR
22	Nozzles/Manways Inspection Report	NIR
23	Vacuum Box Inspection Report	VBR
24	Pneumatic Test Report (Reinforcement Pads)	RPT
25	Dimensional Inspection Report	DIR
26	Platforms and Ladders Inspection Report	PLR
27	Other Attachments Inspection Report	AIR
28	Wind Girder & Roof Structure Inspection Report	RSR
29	Weld Completion Checklist	WCC
30	PWHT Completion Checklist	HTC
31	NDT Completion Checklist	NDC
32	Hydrostatic Test Report	HTR
33	Settlement Measurement Report	SMR
34	Pneumatic Test Report (Roof)	RAT
35	Punch List	PL
36	Compliance Certificate	CC

While these formats are required in the construction of any normal fixed roof tanks, floating roof tanks may require a few more formats, which are indicated in the following table.

15.3 Additional Formats for Floating Roof Tanks

Serial No.	Description	Document Code
37	Annular Clearance Report	ACR
38	Weld Visual Inspection Report (Floating Deck)	VFR
39	NDT Report (Floating Deck)	NDR
40	Rain Water Accumulation Test Report	RWT
41	Roof Flotation Test Report	FTR
42	Pontoon Puncture Test Report	PTR

Though 42 formats are provided in this chapter, all of them may not be required in all cases. Moreover, formats may require additions or deletions to make them ideal for the situation based on the tank size, the complexity in design, and the customer requirements. Therefore, careful reading through code and client specifications is also recommended prior to developing formats for any particular tank.

Logo	**Request for Inspection**	Ref: XXXX-RFI-001 Page: 1 of 1

Project Name	
Tank Name	
Tank No	

From		To	
QA/QC Mgr. (Sub-Con.)		Client	Thru proper channel
Signature			

Notification Date		Inspection Date	
Notification Time		Inspection Time	

☐ Civil ☐ Piping ☐ Pipeline ☐ Mechanical ☐ Electrical ☐ Instrumentation

Location	

ITP Reference		Drawing Reference	
Activity reference		Specification reference	

Inspect the following: ☐ WITNESS ☐ HOLD

Remarks (Sub-Contractor)	**Remarks (Contractor)**	**Remarks (Client)**

Note:
(1) To be issued for all "W" and "H" points in ITP with required notification time as per contract

Sub Contractor		**Contractor**		**Client**	
Name		Name		Name	
Signature		Signature		Signature	
Date		Date		Date	

Logo	Foundation Level-Inspection Report	Ref: XXXX-FLR-001 Page: XX of YY

Project Name: Tank Name: Tank No.:	

Along Circumference Over D1			Along Circumference Over D2			Along Circumference Over D3		
Point	Dim.	Diff.	Point	Dim.	Diff.	Point	Dim.	Diff.
11.25°			22.50°			45.00°		
22.50°			45.00°			90.00°		
33.75°			67.50°			135.00°		
45.00°			90.00°			180.00°		
56.25°			112.50°			225.00°		
67.50°			135.00°			270.00°		
78.75°			157.50°			315.00°		
90.00°			180.00°			360.00°		
101.25°			202.50°					
112.50°			225.00°					
123.50°			247.50°					
135.00°			270.00°					
146.25°			292.50°					
157.50°			315.00°					
168.75°			337.50°					
180.00°			360.00°					
191.25°								
202.50°								
213.75°								
225.00°								
236.25°								
247.50°								
258.75°								
270.00°								
281.25°								
292.50°								
303.75°								
315.00°								
326.25°								
337.50°								
348.75°								
360.00°								

D1
D2
D3

Notes:

(1) Circumference over D1 = Circumference @ outside edge of foundation
(2) Circumference over D2 = Circumference @ 2X D1/3 of foundation
(3) Circumference over D3 = Circumference @ D1/3 of foundation

Sub Contractor		Contractor		Client	
Name		Name		Name	
Signature		Signature		Signature	
Date		Date		Date	

Logo	Material Inspection Report	Ref: XXXX-MIR-001 Page: XX of YY

Project Name	
Tank Name	
Tank No	

Tank No.		Drawing No.	
Purchase Order		Delivery Note.	

Serial No.	Material Description	Material Specification	Size	Qty.	Heat No.	MTC No.	Remarks

Remarks (Sub-Contractor)	Remarks (Contractor)	Remarks (Client)

Note:

(1) To be prepared for all incoming raw materials and shall be referred against each part number in applicable drawing

Sub Contractor		Contractor		Client	
Name		Name		Name	
Signature		Signature		Signature	
Date		Date		Date	

Logo	Welding Consumable Inspection Report	Ref: XXXX-CIR-001 Page: XX of YY

Project Name	
Tank Name	
Tank No	

Purchase Order		Delivery Note.	

Serial No.	Product Details	Brand	Size	Qty.	Lot No.	Batch Test Report	Remarks

Note:

(1) To be prepared for all types of welding consumables like electrodes, filler wires, fluxes, etc. Shielding/trailing gases may be exempted

Sub Contractor		Contractor		Client	
Name		Name		Name	
Signature		Signature		Signature	
Date		Date		Date	

Logo	Welding Consumable Control Log	Ref: XXXX-CCL-001 Page: XX of YY

Project Name	
Tank Name	
Tank No	

Serial No.	Date	Electrode/ Flux	Size	Qty.	Lot/Batch No.	Baking Temp.	Baking Period	Remarks

Note:

(1) To be prepared for all types of welding consumables like electrodes, filler wires & fluxes.

(2) Shielding/trailing gases may be exempted.

Sub Contractor		Contractor		Client	
Name		Name		Name	
Signature		Signature		Signature	
Date		Date		Date	

Logo	Material Traceability Report or Material Summary	Ref: XXXX-MTR-001 Page: XX of YY

Project Name	
Tank Name	
Tank No	

Tank No.		Drawing No.	

Serial No.	Part No.	Material Specification	Heat No.	MTC No.	MIR Reference	Remarks

Note:
(1) To be prepared for all pressure containing parts against each part number of each drawing

Sub Contractor		Contractor		Client	
Name		Name		Name	
Signature		Signature		Signature	
Date		Date		Date	

Logo	Shell Plate Bending Inspection Report	Ref: XXXX-PBR-001 Page: XX of YY

Project Name	
Tank Name	
Tank No	

Tank No.		Drawing No.	
Weld Map Ref:			

Serial No.	Shell Plate Identification	Material Specification	Plate Thickness	Condition or Surface & Edge Preparation	Remarks

Note:

(1) All shell plates including closure plates shall be included in this report

Sub Contractor		Contractor		Client	
Name		Name		Name	
Signature		Signature		Signature	
Date		Date		Date	

Logo	Other Components Bending/Forming Inspection Report	Ref: XXXX-OCR-001 Page: XX of YY

Project Name	
Tank Name	
Tank No	

Applicable Drawings

Drawing No.		Drawing No.	
Drawing No.		Drawing No.	

Serial No.	Component	Material Specifica-tion	Plate Thick-ness	Dimension (Dia & length)		Remarks
				Required	Actual	
1	Sump Neck					
2	Manway Neck M1 (S)					
3	Manway Neck M2 (S)					
4	Manway Neck M3 (S)					
5	Nozzle Neck N1					
6	Nozzle Neck N2					
7	Nozzle Neck N3					
8	Nozzle Neck N4					
9	Clean-out door (parts)					
10	Dome (if applicable)					
11	Pulled bends (if required)					

Notes:

(1) All components requiring cold forming (other than shell plates) shell be included in this report

(2) Items shall be added according to specific requirements in each case

Sub Contractor		Contractor		Client	
Name		Name		Name	
Signature		Signature		Signature	
Date		Date		Date	

Logo	Fit-up Inspection Report	Ref: XXXX-FIR-001 Page: XX of YY

Project Name	
Tank Name	
Tank No	

Tank No.		Drawing No.	
Weld Map Ref:			

Serial No.	Drawing No./Description	Joint No.	Joint Type	Material Spec	WPS No.	Remarks

Note:
(1) To be prepared for all joints identified as pressure containing weld including that of reinforcing pads

Sub Contractor		Contractor		Client	
Name		Name		Name	
Signature		Signature		Signature	
Date		Date		Date	

Logo	Weld Visual Inspection Report	Ref: XXXX-VIR-001 Page: XX of YY

Project Name	
Tank Name	
Tank No	

Tank No.		Drawing No.	
Weld Map Ref:			

Serial No.	Drawing No./Description	Joint No.	Joint Type	Welding Process	WPS No.	Remarks

Notes:

(1) To be prepared for all joints identified as pressure containing weld including that of reinforcing pads
(2) For structural attachment welds; checklist as in format WCC-001 would suffice

Sub Contractor		Contractor		Client	
Name		Name		Name	
Signature		Signature		Signature	
Date		Date		Date	

Logo	Radiographic Test Report	Ref: XXXX-RTR-001 Page: XX of YY

Project Name	
Tank Name	
Tank No	

Tank No.		Drawing No.	
Weld Map Ref:		Procedure No.	

Radiographic Details

Source		Film Type		SFD	
Source Strength		Film Brand		Density	
Source Size		IQI		Sensitivity	
Screen		Technique	☐ SWSI	☐ DWSI	☐ DWDI
Material		Weld Condition	☐ As Welded	☐ After PWHT	

Description of Welds Radiographed

Serial No.	Joint No.	Identification	Thickness	Welder No.	Interpretation	Location	Evaluation	Remarks

Film Size	100×240 mm		100×400 mm	100×200 mm		Others	

Radiographer		Processor		Interpreter	

Abbreviations for Evaluation	Acc.	Acceptable	Rep.	Repair	RT	Re-take

Interpretation Codes		EUC	External Under Cut	RC	Root Concavity
NSD	No Significant Defects	ICP	Incomplete Penetration	IP	Isolated Porosity
ISI	Isolated Slag Inclusion	BT	Burn Through	CP	Cluster Porosity
ESI	Elongated Slag Inclusion	IF	Incomplete Fusion	HB	Hollow Bead Porosity
IRP	Inadequate Root Penetration	TI	Tungsten Inclusion	LF	Lack of Fusion
IUC	Internal Under Cut	EP	Excess Penetration	CR	Crack

Note:

(1) Every segment radiographed shall be covered in this report

Sub Contractor		Contractor		Client	
Name		Name		Name	
Signature		Signature		Signature	
Date		Date		Date	

Logo	Ultrasonic Test Report	Ref: XXXX-UTR-001 Page: XX of YY

Project Name	
Tank Name	
Tank No	

Tank No.		Drawing No.	
Weld Map Ref:		Procedure No.	

Ultrasonic Test Details

Equipment Type		Area Scanned		Time Base Range	
Model/Sl. No.		Surface Condition		Defect Report Level	
Certificate No.		Couplant		Defect Reject Level	
Calibration Block		Transfer Correction		Stage	Before/After PWHT
Material & Thick-ness	-	Sensitivity	Reference dB		
			Scanning dB	-	

Details of Probes

Angle	Frequency	Size	Type	Angle	Frequency	Size	Type
(1)				(2)			
(3)				(4)			
(5)				(6)			

Description of Welds Tested

Serial No.	Joint No.	Location	Probe < Indication	Defect Type	Defect Size (mm)		Reference Height (+dB)	Evaluation
					Length	Depth		

Test Results & Comments (if any)	

Technician		Witnessed by		Interpreter	

Notes:
(1) Every segment ultrasonically tested shall be clearly covered by this report
(2) Additional sketches (if required) also shall be attached to describe methodology

Sub Contractor		**Contractor**		**Client**	
Name		Name		Name	
Signature		Signature		Signature	
Date		Date		Date	

Logo	Magnetic Particle Test Report	Ref: XXXX-MPR-001 Page: XX of YY

Project Name	
Tank Name	
Tank No	

Tank No.		Drawing No.	
Weld Map Ref:		Procedure No.	

Magnetic Particle Test Details

Equipment Type		Magnetic Particle	
Model/Sl. No.		Material	
Certificate No. & Validity		Surface Condition	
Contrast Paint		State	Before/After PWHT
Method	☐ Wet ☐ Dry ☐ Fluorescent	Lighting	☐ Natural ☐ Artificial ☐ Ultraviolet
-			

Description of Welds Tested

Serial No.	Joint No.	Location	Indication	Defect Size	Evaluation

Test Results & Comments (if any)	

Technician		Witnessed by		Interpreter	

Notes:
(1) Every segment tested shall be clearly covered by this report
(2) Additional sketches (if required) also shall be attached to describe methodology

Sub Contractor		**Contractor**		**Client**	
Name		Name		Name	
Signature		Signature		Signature	
Date		Date		Date	

Logo	Dye Penetrant Test Report	Ref: XXXX-DPR-001 Page: XX of YY

Project Name	
Tank Name	
Tank No	

Tank No.		Drawing No.	
Weld Map Ref:		Procedure No.	

Dye Penetrant Test Details

Equipment Type		Cleaner Type	
Material		Cleaner Application	
Surface Condition		Developer Type	
Penetrant Type		Development Time	
Dwell Time		Stage	Before/After PWHT
Method	☐ Color dye ☐ Fluorescent dye	Lighting	☐ Natural ☐ Artificial ☐ Ultraviolet

Description of Welds Tested

Serial No.	Joint No.	Location	Indication	Defect Size	Evaluation

Test Results & Comments (if any)	

Technician		Witnessed by		Interpreter	

Note:
(1) Every segment of weld or entire weld tested shall be clearly covered by this report

Sub Contractor		**Contractor**		**Client**	
Name		Name		Name	
Signature		Signature		Signature	
Date		Date		Date	

Logo	Post Weld Heat Treatment Report	Ref: XXXX-PWH-001 Page : XX of YY

Project Name	
Tank Name	
Tank No	

Tank No.		Drawing No.	
Weld Map Ref:		Procedure No.	

PWHT Cycle Details

Heating Rate		Soaking Temperature		Soaking Time	
Cooling Rate		Loading Temperature		Cooling below 425 °C	Still air

Details of Recorder

Equipment No.		Type		Calibration	

Details of Thermocouples & Compensating Cables

Type		Composition		Cable	
Number		Locations			

Actual Parameters of PWHT

Heating Rate		Soaking Temperature		Soaking Time	
Cooling Rate		Loading Temperature		Chart Speed	

Items Post Weld Heat Treated

Sl. No	Description	Remarks

Special Observations or Comments(if any)	

Technician		Witnessed by		QC Engineer	

Notes:

(1) Heating & Cooling rates specified are above 425°C (800°F)

(2) Location of thermocouples may be indicated through sketches (if required)

Sub Contractor		Contractor		Client	
Name		Name		Name	
Signature		Signature		Signature	
Date		Date		Date	

Logo	Hardness Survey Report	Ref: XXXX-HSR-001 Page: XX of YY

Project Name	
Tank Name	
Tank No	

Tank No.		Drawing No.	
Weld Map Ref:		Procedure No.	

Hardness Test Details

Equipment Type	
Model/Sl. No.	
Certificate No & Validity	
Material	
Stage	Before/After PWHT

A B C D E

Description of Welds Tested

Serial No.	Joint No.	Location & Hardness Values (BHN)					Remarks
		PM(A)	HAZ(B)	Weld (C)	HAZ(D)	PM(E)	

Test Results & Comments (if any)	

Technician		Witnessed by		Interpreter	

Notes:

(1) Every segment tested shall be clearly covered by this report

(2) Additional sketches (if required) also shall be attached to describe methodology

Sub Contractor		**Contractor**		**Client**	
Name		Name		Name	
Signature		Signature		Signature	
Date		Date		Date	

Logo	Weld Summary Report	Ref: XXXX-WSR-001 Page: XX of YY

Project Name	
Tank Name	
Tank No	

Tank No.		Drawing No.	
Weld Map Ref:			

Serial No.	Weld Joint No.	Joint Type	Fit-up	WPS	Welder	Visual	RT/UT	MPT	VBT	Remarks

Notes:

(1) All joints identified as pressure containing welds including that of reinforcing pads to be covered in this report

(2) Report reference numbers shall be provided against each inspection/test to ensure completeness

Sub Contractor		Contractor		Client	
Name		Name		Name	
Signature		Signature		Signature	
Date		Date		Date	

Logo	Weld Peaking Control Report	Ref: XXXX-PCR-001 Page: XX of YY

Project Name	
Tank Name	
Tank No	

Tank No.		Drawing No.	
Weld Map Ref:			

Serial No.	Shell Course	Joint No.	Peaking @ Fit-up			Peaking after Welding			Remarks
			Top	Middle	Bottom	Top	Middle	Bottom	

Notes:

(1) Peaking at all vertical shell joints shall be recorded

(2) Use either inside or outside profile gauge of approximately 2.5 m chord length or more for measuring peaking

Sub Contractor		Contractor		Client	
Name		Name		Name	
Signature		Signature		Signature	
Date		Date		Date	

Logo	Banding & Circumference Report	Ref: XXXX-BCR-001 Page: XX of YY

Project Name	
Tank Name	
Tank No	

Tank No.		Drawing No.	
Weld Map Ref:			

Serial No.	Shell Course	Joint Nos.	Circumference at Fit-up			Circumference after Welding			Remarks
			Top	Middle	Bottom	Top	Middle	Bottom	

Notes:

(1) Top & bottom measurements per shell course shall be taken at within 25 mm from edge of weld on either side of weld for all horizontal seams

(2) While taking measurements, ensure that measuring tape is placed horizontally in a straight line

Sub Contractor		Contractor		Client	
Name		Name		Name	
Signature		Signature		Signature	
Date		Date		Date	

Logo	Plumbness Inspection Report	Ref: XXXX-PIR-001 Page: XX of YY

Project Name	
Tank Name	
Tank No	

Tank No.		Drawing No.	
Weld Map Ref:			

Serial No.	Shell No.	Location	Plumb (Verticality)								Remarks
			45°	90°	135°	180°	225°	270°	315°	360°	
1	Shell 1	Bottom									
2		Middle									
3		Top									
4	Shell 2	Bottom									
5		Middle									
6		Top									
7	Shell 3	Bottom									
8		Middle									
9		Top									
10	Shell 4	Bottom									
11		Middle									
12		Top									
13	Shell 5	Bottom									
14		Middle									
15		Top									
16	Shell 6	Bottom									
17		Middle									
18		Top									
19	Shell 7	Bottom									
20		Middle									
21		Top									
22	Shell 8	Bottom									
23		Middle									
24		Top									

Notes:

(1) Measurements shall be taken for all shell courses as in report format

(2) The tank cited as example in this book has 8 shell courses and hence format is developed for 8 shell courses

Sub Contractor		Contractor		Client	
Name		Name		Name	
Signature		Signature		Signature	
Date		Date		Date	

Logo	Roundness Inspection Report	Ref: XXXX-RIR-001 Page: XX of YY

Project Name	
Tank Name	
Tank No	

Tank No.		Drawing No.	
Weld Map Ref:			

Serial No.	Shell No.	Location	Roundness (inside diameter)								Remarks
			45°	90°	135°	180°	225°	270°	315°	360°	
1	Shell 1	Bottom									
2		Middle									
3		Top									
4	Shell 2	Bottom									
5		Middle									
6		Top									
7	Shell 3	Bottom									
8		Middle									
9		Top									
10	Shell 4	Bottom									
11		Middle									
12		Top									
13	Shell 5	Bottom									
14		Middle									
15		Top									
16	Shell 6	Bottom									
17		Middle									
18		Top									
19	Shell 7	Bottom									
20		Middle									
21		Top									
22	Shell 8	Bottom									
23		Middle									
24		Top									

Notes:

(1) Measurements shall be taken for all shell courses as in report format

(2) The tank cited as example in this book has 8 shell courses and hence format is developed for 8 shell courses

Sub Contractor		Contractor		Client	
Name		Name		Name	
Signature		Signature		Signature	
Date		Date		Date	

Logo	Nozzles/Manways Inspection Report	Ref: XXXX-NIR-001 Page: XX of YY

Project Name	
Tank Name	
Tank No	

Tank No.		Drawing No.	

Serial No.	Nozzle/Manway Identification	Size	Elevation		Orientation		Remarks
			Required	Actual	Required	Actual	

Notes:

(1) To be prepared for all Nozzles and Manways

(2) If required same format can be used for other attachments also

Sub Contractor		Contractor		Client	
Name		Name		Name	
Signature		Signature		Signature	
Date		Date		Date	

Logo	Vacuum Box Inspection Report	Ref: XXXX-VBR-001 Page: XX of YY

Project Name	
Tank Name	
Tank No	

Tank No.		Drawing No.	
Weld Map Ref:		Procedure No.	

Vacuum Gauge Details

Identification		Certificate No.	
Make		Validity	
Range		Bubble Solution	
Date of Calibration		Metal Temperature	-

Test Pressure	

Serial No.	Joint No.	Thickness	Length	Observation	Remarks

Note:
(1) Welds joint numbers listed shall be correlated to respective weld maps

Sub Contractor		Contractor		Client	
Name		Name		Name	
Signature		Signature		Signature	
Date		Date		Date	

Logo	Pneumatic Test Report (Reinforcement Pads)	Ref: XXXX-RPT-001 Page: XX of YY

Project Name	
Tank Name	
Tank No	

Tank No.		Drawing No.	
Weld Map Ref:		Procedure No.	

Pressure Gauge Details

Identification		Certificate No.	
Make		Validity	
Range		Bubble Solution	
Date of Calibration	-	Metal Temperature	

Test Pressure	

Serial No.	Nozzle/Manway Identification	Joint No.	Welder No.	Remarks

Note:

(1) Nozzle identifications of all nozzles shall be correlated to respective drawings and weld maps

Sub Contractor		Contractor		Client	
Name		Name		Name	
Signature		Signature		Signature	
Date		Date		Date	

Logo	Dimensional Inspection Report	Ref: XXXX-DIR-001 Page: XX of YY

Project Name	
Tank Name	
Tank No	

Tank No.		Drawing No.	

Serial No.	Description of Dimension	Identification (if any)	Dimension		Remarks
			Required	Actual	

Note:
(1) All salient dimensions shown in drawings shall be reported in this format

Sub Contractor		Contractor		Client	
Name		Name		Name	
Signature		Signature		Signature	
Date		Date		Date	

Logo	Platform & Ladder Inspection Report	Ref: XXXX-PLR-001 Page: XX of YY

Project Name	
Tank Name	
Tank No	

Tank No.		Drawing No.	

Serial No.	Part Name	Position or Identification	Item Inspected	Welding Status	Remarks
1	Stairway	External	Supports		
2			Grating		
3			Handrail		
4			Intermediate Platform		
5			Bolting if any		
6	Ladder	Internal	Supports	N/A	
7			Rungs		
8			Cage		
9			Intermediate landing		
10	Roof Platforms	Access Platforms and walk ways	Supports		
11			Platform Structure		
12			Handrails		
13	Roof Handrails	On roof	Roof Handrail Posts		
14			Handrails		
15			Intermediate brazing		
16	Lighting Poles	Mountings	Poles		
17	Instrumentation Accessories	Mountings	Attachment welds		
18	Lightning arrestor	Mountings	Welding/ Clamping		

Notes:

(1) To be prepared for each flight of stairway and ladder in similar way

(2) List out any other accessory which is welded to tank roof if not covered by other reports

Sub Contractor		Contractor		Client	
Name		Name		Name	
Signature		Signature		Signature	
Date		Date		Date	

Logo	Other Attachments Inspection Report	Ref: XXXX-AIR-001 Page: XX of YY

Project Name	
Tank Name	
Tank No	

Tank No.		Drawing No.	

Serial No.	Part Name	Position or Identification	Number Required	Number Inspected	Remarks
		External			
1	Pipe Support Cleats	Shell			
2	Cooling Rings	Shell			
3	Instrumentation mounting cleats	Shell			
4	Fire Water Pipe supports	Shell/roof			
5	F&G system supports	Shell/roof			
		Internal			
6	Caged ladder Cleats	Shell			
7	Distributor Pipe supports	Shell/Bottom Plate			
8	Draw of Pipe supports	Bottom Plate			
9	Vortex Breakers	Shell			

Notes:
(1) List out any other accessory which is welded to tank shell/roof if not covered by other reports

Sub Contractor		Contractor		Client	
Name		Name		Name	
Signature		Signature		Signature	
Date		Date		Date	

Logo	Wind Girder, Roof Structure Inspection Report	Ref: XXXX-RSR-001 Page: XX of YY

Project Name	
Tank Name	
Tank No	

Tank No.		Drawing No.	

Serial No.	Part Name	Position	Welding	Bolting	Remarks
1	Compression Ring	Shell to CR		N/A	
2	Curb Angle	Shell to CA		N/A	
3	Center Drum	Weld in central drum		N/A	
4		R S support		N/A	
5	Roof Structure	Weld within		N/A	
6		Welds with brazing		N/A	
7		To Shell support	N/A		
8		To central drum support	N/A		
9	Wind Girder	To shell		N/A	

Notes:

(1) To be prepared for all types of strengthening attachments wherein bolting is also vital.

(2) Observe for discrepancies with regard to shortage of fastners/welding.

(3) Welding fillet sizes shall be optimal.

Sub Contractor		Contractor		Client	
Name		Name		Name	
Signature		Signature		Signature	
Date		Date		Date	

Logo	Weld Completion Checklist	Ref: XXXX-WCC-001 Page: XX of YY

Project Name	
Tank Name	
Tank No	

Tank No.		Drawing No.	

Serial No.	Location	Description	Weld Status	Remarks
1	Bottom Plate	Annular butt		
2		Fillets		
3		Fillet BP to sketch plate		
4		Sump		
5	Shell	To annular		
6		Vertical		
7		Horizontal		
8		To crub angle		
9		To compression ring		
10		Wind girders		
11		Nozzles		
12		Manways		
13		Clean-out doors		
14		Internals		
15		Structural supports		
16		Internal pipe supports		
17		Other attachments		
18	Roof	Roof to CA		
19		Roof to CR		
20		Fillet of roof plate		
21		Nozzles		
22		Manways		
23		Pipe & other supports		
24	Stairways	To shell		
25		Welds within		
26	Platforms	To shell		
27		To roof		
28		Welds within		
29	Others	Shell		
30		Roof		
31	Center Drum	Weld within		
32	Roof Structure	Weld within		
33	Internal supports	To shell/BP		
34	Internals	Welds within		

Note:
(1) This is just a consolidation to be prepared based on individual reports for each item

Sub Contractor		Contractor		Client	
Name		Name		Name	
Signature		Signature		Signature	
Date		Date		Date	

Logo	PWHT Completion Checklist	Ref: XXXX-HTC-001 Page: XX of YY

Project Name	
Tank Name	
Tank No	

Tank No.		Drawing No.	

Serial No.	Component Description	Type of Heat Treatment	Soaking Time	Soaking Temperature	Report Reference	Remarks

Notes:

(1) This is a consolidation of details, prepared based on individual reports for each item requiring PWHT or any other heat treatment other than that for bought out items

Sub Contractor		Contractor		Client	
Name		Name		Name	
Signature		Signature		Signature	
Date		Date		Date	

Logo	NDT Completion Checklist	Ref: XXXX-NDC-001 Page: XX of YY

Project Name	
Tank Name	
Tank No	

Tank No.		Drawing No.	

Serial No.	Location	Description	Weld Status				Remarks
	Bottom Plate	Annular butt					
		Fillets					
		Fillet BP to sketch plate					
		Sump					
	Shell	To annular					
		Vertical					
		Horizontal					
		To crub angle					
		To compression ring					
		Wind girders					
		Nozzles					
		Manways					
		Clean-out doors					
		Internals					
		Structural supports					
		Internal pipe supports					
		Other attachments					
	Roof	Roof to CA					
		Roof to CR					
		Fillet of roof plate					
		Nozzles					
		Manways					
		Pipe & other supports					
	Stairways	To shell					

Sub Contractor		Contractor		Client	
Name		Name		Name	
Signature		Signature		Signature	
Date		Date		Date	

Logo	NDT Completion Check List	Ref: XXXX-NDC-001 Page: XX of YY

		Welds within					
	Platforms	To shell					
		To roof					
		Welds within					
	Others	Shell					
		Roof					
	Center Drum	Weld within					
	Roof Structure	Weld within					
	Internal supports	To shell/BP					
	Internals	Welds within					

Note:

(1) This is just a consolidation to be prepared based on individual reports for each item

Sub Contractor		Contractor		Client	
Name		Name		Name	
Signature		Signature		Signature	
Date		Date		Date	

Logo	Hydrostatic Test Report	Ref: XXXX-HTR-001 Page: XX of YY

Project Name	
Tank Name	
Tank No	

Tank No.		Drawing No.	
Weld Map Ref:		Procedure No.	

Manometer Details (If used for Roof Air Test)

Identification		Certification No.	
Make		Calibration Validity	
Range		Metal Temperature	

Details of Test Medium

Test Medium	
Inhibitors (if any)	
Dosing Rate	

Filling					
Serial No.	**Fill Level**	**Date**	**Stabilization**		**Evaluation**
			Start	**Finish**	
1	Filling Start				
2	25% Test Height				
3	50% Test Height				
4	75% Test Height				
5	100% Test Height				
Emptying					
6	50% Test Height				
7	Empty				
Observations					

Notes:

(1) Calibration Reports (if any) shall be attached with this report.
(2) Test report for medium used also shall be enclosed.
(3) Dosing details with its material safety data sheet (MSDS) also shall be enclosed.
(4) Indicate the draining & drying process adopted after hydrostatic test prior to surface preparation and lining under observations.

Sub Contractor		**Contractor**		**Client**	
Name		Name		Name	
Signature		Signature		Signature	
Date		Date		Date	

Logo	Settlement Measurement Report	Ref: XXXX-SMR-001 Page: XX of YY

Project Name: Tank Name: Tank No.:	

Tank Settlement Readings

Serial No.	Angle	Filling									Emptying		
		Empty	25%	25%	50%	50%	75%	75%	100%	100%	50%	50%	Empty
1	11.25°												
2	22.50°												
3	33.75°												
4	45.00°												
5	56.25°												
6	67.50°												
7	78.75°												
8	90.00°												
9	101.25°												
10	112.50°												
11	123.50°												
12	135.00°												
13	146.25°												
14	157.50°												
15	168.75°												
16	180.00°												
17	191.25°												
18	202.50°												
19	213.75°												
20	225.00°												
21	236.25°												
22	247.50°												
23	258.75°												
24	270.00°												
25	281.25°												
26	292.50°												
27	303.75°												
28	315.00°												
29	326.25°												
30	337.50°												
31	348.75°												
32	360.00°												
Remarks													

Notes:

(1) Depending on circumference of tank, number of readings may increase or decrease

(2) Measurements required on attaining level & after stabilization time as specified by code/client

(3) Stabilization time 24 hrs for all stages (API 650). (48 hrs for 100% full by clients in Oil & Gas)

Sub Contractor		Contractor		Client	
Name		Name		Name	
Signature		Signature		Signature	
Date		Date		Date	

Logo	Pneumatic Test Report (Roof)	Ref: XXXX-RAT-001 Page: XX of YY

Project Name	
Tank Name	
Tank No	

Tank No.		Drawing No.	
Weld Map Ref:		Procedure No.	

Pressure Gauge/Manometer Details

Identification		Certificate No.	
Make		Validity	
Range		Bubble Solution	
Date of Calibration		Metal Temperature	

Test Pressure	

Serial No.	Details of Roof Welds	Joint No.	Welder No.	Remarks

Note:

(1) All weld joints covered by this test shall be included in the report & shall be correlated to weld map

Sub Contractor		Contractor		Client	
Name		Name		Name	
Signature		Signature		Signature	
Date		Date		Date	

Logo	Punch List	Ref: XXXX-PL-001 Page: XX of YY

Project Name	
Tank Name	
Tank No	

Tank No.		G A Drawing No.	

Serial No.	Drawing No	Description of Outstanding Work	Category of Punch			Remarks
			A	B	C	

Notes:

(1) To be prepared against each detail drawing to make it comprehensive enough
(2) Punch items need to be categorized as A, B, and C depending on severity of each
(3) Category "A" is for items that are to be carried out before Hydrostatic Testing
(4) Category "B" is for items that can be carried even after Hydrostatic Testing, but before surface preparation & painting
(5) Category "C" is for items that can be carried even after painting but before commissioning, without affecting any of the completed works, like replacement of a valve or a fitting, etc.

Sub Contractor		Contractor		Client	
Name		Name		Name	
Signature		Signature		Signature	
Date		Date		Date	

Logo	Manufacturer's Certification for Tanks Built to API Standard 650	Ref: XXXX-CC-001 Page: XX of YY

Project Name	
Tank Name	
Tank No	
Tank Size & Capacity	
Tank Type	Fixed/Floating Roof

Serial No.	Drawing No.	Description

Serial No.	Drawing No.	Description

This is to certify that the tank constructed for --

----
at (location) -- according to drawings and documents mentioned above meets all applicable requirements of API Standard 650, ----------------Edition,----------------Revision, Annex---------------, including the requirements for design, materials, fabrication, and erection.

The tank is further described on the attached as-built data sheet No---------- Rev -------- dated -----------

Contractor/Manufacturer (Address)	
Authorized Representative	(Name)
Signature	Stamp
Date	

Notes:
(1) List out all applicable Drawings and Procedures used in construction in annexure if required.
(2) As-built data sheet also shall be enclosed as an attachment.

Logo	Annular Clearance Report	Ref: XXXX-ACR-001 Page: XX of YY

Project Name	
Tank Name	
Tank No	

Tank No.		Drawing No.	

Serial No.	**Shell No.**	**Location**	**Annular Clearance**								**Remarks**
			45°	**90°**	**135°**	**180°**	**225°**	**270°**	**315°**	**360°**	
1	Shell 1	Bottom									
2		Middle									
3		Top									
4	Shell 2	Bottom									
5		Middle									
6		Top									
7	Shell 3	Bottom									
8		Middle									
9		Top									
10	Shell 4	Bottom									
11		Middle									
12		Top									
13	Shell 5	Bottom									
14		Middle									
15		Top									
16	Shell 6	Bottom									
17		Middle									
18		Top									
19	Shell 7	Bottom									
20		Middle									
21		Top									
22	Shell 8	Bottom									
23		Middle									
24		Top									

Notes:

(1) Measurements shall be taken for all shell courses as in report format

(2) Depending on lowest position of floating deck, it may not be possible to record clearance in shell course (1)

(3) The tank cited as example in this book has 8 shell courses and hence format is developed for 8 shell courses

Sub Contractor		**Contractor**		**Client**	
Name		Name		Name	
Signature		Signature		Signature	
Date		Date		Date	

Logo	Weld Visual Inspection Report (Floating Deck)	Ref: XXXX-VFR-001 Page: XX of YY

Project Name	
Tank Name	
Tank No	

Tank No.		Drawing No.	
Weld Map Ref:			

Serial No.	Drawing No./Description	Joint No.	Joint Type	Welding Process	WPS No.	Remarks

Notes:

(1) To be prepared for all joints identified on floating deck including that of openings passing through pontoon or deck plate or compartments (in double deck type)

(2) Welds of stiffeners and pads also shall be included in this report as a summary (if not listed in detail)

Sub Contractor		Contractor		Client	
Name		Name		Name	
Signature		Signature		Signature	
Date		Date		Date	

Logo	NDT Report (Floating Deck)	Ref: XXXX-NDR-001 Page: XX of YY

Project Name	
Tank Name	
Tank No	

Tank No.		Drawing No.	
Weld Map Ref:		Procedure No.	

Dye Penetrant Test

Equipment Type		Cleaner Type	
Material		Cleaner Application	
Surface Condition		Developer Type	
Penetrant Type		Development Time	
Dwell Time		Stage	
Method ☐ Color dye ☐ Fluorescent dye		Lighting	☐ Natural ☐ Artificial ☐ Ultraviolet

Description of Welds Tested

Serial No.	Joint No.	Location	Indication	Defect Size	Evaluation

Test Results & Comments (if any)	

Technician		Witnessed by		Interpreter	

Chalk Oil Test

Oil Used		Chalk Application	
Application		Dwell Time	
Mental Temperature		Posttest Cleaning	

Description of Welds Tested

Serial No.	Joint No.	Location	Indication	Defect Size	Evaluation

Sub Contractor		Contractor		Client	
Name		Name		Name	
Signature		Signature		Signature	
Date		Date		Date	

Logo	NDT Report (Floating Deck)	Ref: XXXX-NDR-001 Page: XX of YY

Test Results & Comments (if any)	

Technician		Witnessed by		Interpreter	

Vacuum Box Test

Identification		Certificate No.	
Make		Validity	
Range		Bubble Solution	
Date of Calibration		Mental Temperature	

Test Pressure	

Serial No.	**Joint No.**	**Thickness**	**Length**	**Observation**	**Remarks**

Notes:

(1) All the tree tests indicated above may not be necessary for all the welds

(2) Client specifications often require more than one test for atleast a few critical welds on floating deck

(3) Portions of this format (if not applicable) may be deleted before implementing the same in any project

Sub Contractor		**Contractor**		**Client**	
Name		Name		Name	
Signature		Signature		Signature	
Date		Date		Date	

Logo	Rainwater Accumulation Test Report	Ref: XXXX-RWT-001 Page: XX of YY

Project Name	
Tank Name	
Tank No	

Tank No.		Drawing No.	
Weld Map Ref:		Procedure No.	

Rainwater Accumulation Test Conditions

Pontoon Intact & with Full Height Rainwater Accumulation

Test Medium	Potable Water	NRV Condition	Closed
Tank Fill Level	100%	Primary Roof Drain	Open at tank shell
Rainwater Fill Level	100%	Pontoon Condition	All intact
Holding Time	24 hours	Stabilization	Satisfactory

Serial No.	Immersion								Remarks
	45°	90°	135°	180°	225°	270°	315°	360°	
1									@ Start of holding
2									@ End of Holding

One Pontoon Punctured & with Full Height Rainwater Accumulation

Test Medium	Potable Water	NRV Condition	Closed
Tank Fill Level	100%	Primary Roof Drain	Open at tank shell
Rainwater Fill Level	100%	Pontoon Condition	**One Pontoon Punctured**
Holding Time	24 hours	Stabilization	Satisfactory

Serial No.	Immersion								Remarks
	45°	90°	135°	180°	225°	270°	315°	360°	
3									@ Start of holding
4									@ End of Holding

Two Pontoon Punctured & with Full Height Rainwater Accumulation

Test Medium	Potable Water	NRV Condition	Closed
Tank Fill Level	100%	Primary Roof Drain	Open at tank shell
Rainwater Fill Level	100%	Pontoon Condition	**Two Pontoons Punctured**
Holding Time	24 hours	Stabilization	Satisfactory

Serial No.	Immersion								Remarks
	45°	90°	135°	180°	225°	270°	315°	360°	
5									@ Start of holding
6									@ End of Holding

Note:

(1) At each stage, before taking measurements, it shall be ensured that the floating deck is stabilized in that position

Sub Contractor		Contractor		Client	
Name		Name		Name	
Signature		Signature		Signature	
Date		Date		Date	

Logo	Roof Floatation Test Report	Ref: XXXX-FTR-001 Page: XX of YY

Project Name	
Tank Name	
Tank No	

Tank No.		Drawing No.	
Weld Map Ref:		Procedure No.	

Roof Floatation Test Details

Test Medium		NRV Condition	Closed/Open

Serial No.	Water Fill Stage	Annular Clearance								Remarks
		45°	90°	135°	180°	225°	270°	315°	360°	
Filling										
1	Before Fill									
2	25% Full									
3	50% Full									
4	75% Full									
5	100% Full									
Emptying										
6	50% Full									
7	Resting on legs									

Rolling Ladder Position & Level

Serial No.	Water Fill Stage	Ladder Movement (along rail)	Level of Ladder Treads	Remarks
Filling				
1	Before Fill			
2	25% Full			
3	50% Full			
4	75% Full			
5	100% Full			
Emptying				
6	50% Full			
7	Resting on legs			

Floating of Roof	Smooth/Rough	Legs before & after Test	Satisfactory
Leg support positions	Satisfactory	Anti-rotation device	Satisfactory

Notes:

(1) While emptying, the punctured compartments may be filled with water

(2) Because of (1) above, annular clearnace while filling & draining need not be the same

Sub Contractor		Contractor		Client	
Name		Name		Name	
Signature		Signature		Signature	
Date		Date		Date	

Logo	Pontoon Puncture Test Report	Ref: XXXX-PTR-001 Page: XX of YY

Project Name	
Tank Name	
Tank No	

Drawing No.		Procedure No.	

Pontoon Puncture Test Conditions (All Pontoons Intact)

Test Medium	N/A	NRV Condition	Closed
Tank Fill Level	100%	Primary Roof Drain	Open at tank shell
Rainwater Fill Level	Nil	Pontoon Condition	All intact
Holding Time	24 hours	Stabilization	Satisfactory

Serial No.	Annular Clearance								Remarks
	45°	90°	135°	180°	225°	270°	315°	360°	
1									@ Start of holding
2									@ End of holding

One Pontoon Punctured

Test Medium	Potable Water	NRV Condition	Closed
Tank Fill Level	100%	Primary Roof Drain	Open at tank shell
Rainwater Fill Level	Nil	Pontoon Condition	**One Pontoon Punctured**
Holding Time	24 hours	Stabilization	Satisfactory

Serial No.	Annular Clearance								Remarks
	45°	90°	135°	180°	225°	270°	315°	360°	
3									@ Start of holding
4									@ End of holding

Two Pontoons Punctured

Test Medium	Potable Water	NRV Condition	Closed
Tank Fill Level	100%	Primary Roof Drain	Open at tank shell
Rainwater Fill Level	Nil	Pontoon Condition	**Two Pontoons Punctured**
Holding Time	24 hours	Stabilization	Satisfactory

Serial No.	Annular Clearance								Remarks
	45°	90°	135°	180°	225°	270°	315°	360°	
5									@ Start of holding
6									@ End of holding

Two Pontoons Punctured & with Full Height Rainwater on Roof

Test Medium	Potable Water	NRV Condition	Closed
Tank Fill Level	100%	Primary Roof Drain	Open at tank shell
Rainwater Fill Level	Nil	Pontoon Condition	**Two Pontoons Punctured**
Holding Time	24 hours	Stabilization	Satisfactory

Serial No.	Annular Clearance								Remarks
	45°	90°	135°	180°	225°	270°	315°	360°	
7									@ Start of holding
8									@ End of holding

Note:

(1) At each stage, before taking measurements, it shall be ensured that the floating deck is stabilized in its position.

Sub Contractor		Contractor		Client	
Name		Name		Name	
Signature		Signature		Signature	
Date		Date		Date	

Annexure A: Material Specification Summary

API 650 considered a range of commercially used materials for the construction of storage tanks in its Clause 4 under various specifications like ASTM, CSA, ISO, and EN. Clause 4 of API 650 also specifies the special requirements or restrictions on material specifications considered by it. This information is provided against product forms such as plates, sheets, structural shapes, pipes and pipe fittings, forgings, flanges, bolting, welding electrodes, and gaskets—the usual product forms required in construction of a storage tank.

By providing a limited number of specifications in API 650, it does not exclude the use of product forms manufactured according to other national standards. Materials manufactured under other national standards are also acceptable, provided they comply with the requirements specified for the listed material in general and when approved by the end user.

Materials that are usually considered in the construction of storage tanks as per API 650 are summarized as a table. For full details, see Clause 4 of API 650. (Though many more materials are specified in Clause 4, only ASTM materials are provided in this annexure, being in use predominantly.)

The following tables provide lists of material specifications (API, ASTM, and AWS) as a quick guidance under each product considered by API 650 and commonly used worldwide in the industry, except for gaskets. In the case of gaskets, the section provides only certain conditions or restrictions applicable to gaskets. (See API 650 Clause 4 to get the full requirements for all product forms.)

Plates

Group I: A 283 M C, A 285 M C, A 131 M A, A 36 M

Group II: A 131 M B, A 36 M

Group III: A 573 M-400, A 516 M-380, A 516 M-415

Group III A: A 131 C S, A 573 M-400, A 516 M-380, A 516 M-415

Group IV: A 573 M-450, A 573 M-485, A 516 M-450, A 516 M-485, A 662 M B

Group IV A: A 662 M C, A 573 M-485

Group V: A 573 M-485, A 516 M-450, A 516 M-485

Group VI: A 131 M EH 36, A 633 M C, A 633 M D, A 573M C1, A 573 M C 2, A 678 M A, A 678 M B, A 737 M B, A 841 Gr A C 1, A 841 Gr B C 2

Sheets

1. A 1011M, Gr33

(*Continued*)

Pipes

1. API Spec 5L, Grades A, B, and X42
2. ASTM A 53M/A 53, Grades A and B
3. ASTM A 106 M/A 106, Grades A and B; ASTM A 234M/A 234, Grade WPB
4. ASTM A 333M/A 333, Grades 1 and 6
5. ASTM A 334M/A 334, Grades 1 and 6
6. ASTM A 524, Grades I and II
7. ASTM A 671 (see Clause 4.5.3)

Pipe Fittings

1. ASTM A234M/A234, Grade WPB
2. ASTM A 420M/A 420, Grade WPL6

Forgings

1. ASTM A 105M/A 105
2. ASTM A 181M/A 181
3. ASTM A 350M/A 350, LF1 and LF2

Bolting

1. Flange bolting shall be of ASTM A 193 B7, with dimensions as per ASME B18.2.1.
2. Nuts shall be of ASTM A 194 Grade 2H with dimensions as per ASME B18.2.2.
3. Bolts and nuts shall have a heavy hexagonal pattern.
4. Bolts and nuts shall be threaded as per ASME B1.13M or B 1.1.
5. Other bolting (except flange) shall be of ASTM A 307 or A 193M/A 193.
6. A 325M/A 325 is permitted for structural purposes only.

Electrodes

1. For welding of materials with a minimum tensile strength less than 550 MPa (80 ksi), the MMAW electrodes shall be of E60 and E70 classification series (AWS 5.1).
2. For welding of materials with a minimum tensile strength of 550 MPa–585 MPa (80 ksi–85 ksi), MMAW electrodes shall be of E80XX-CX classification series (AWS A5.5).

Gaskets

1. Sheet gaskets shall be continuous.
2. Welded metal gaskets are acceptable subject to conditions.
3. Rope or tape gaskets shall have overlapped ends.
4. Gaskets shall have an integral centering or positioning device.
5. Joint sealing compounds of any sort not allowable except as approved.
6. Material shall be chemically compatible with the fluid stored and the flange.

Annexure B: Recommended Joint Design Guide to Sketches and Tables of API 650

API 650 recommends acceptable typical designs for various types of joints envisaged in tank construction. The user may require slightly different versions of joints suitable for the site conditions, but generally joints shall be in line with the requirements spelled out for typical joints with full compliance to the logic and philosophy behind the recommended design. The following table serves as a guide to figures and tables for each type of weld joint envisaged in the storage tank.

Sl. No.	Description	Figure or Table Reference (API 650)	Remarks
1	Annular, Sketch, and Bottom Plate Weld Joints	Figure 5.3(a)	
2	Annular Plate Thickness Based on Diameter of Tank	Table 5.1(a)	
3	Typical Weld Joints for Vertical Seams in Shell	Figure 5.1	
4	Shell Plate Thickness Based on Diameter of Tank	Table in Clause 5.6.1.1	
5	Typical Weld Joints in Horizontal Seams in Shell	Figure 5.2	
6	Typical Roof Plate and Roof to Curb Angle Joints	Figure 5.3(a)	
7	Minimum Sizes of Top Angles	Table in Clause 5.1.5.9	
8	Typical Joints (Shell to Annular Plate)	Figure 5.3(a)	
9	Typical Weld Joint of Shell to Annular Plate (AP ≥ 13 mm)	Figure 5.3(c)	
10	Fillet Weld Sizes Based on Bottom Shell Course Thickness	Table in Clause 5.1.5.7	
11	Typical Lap Preparation under Shell	Figure 5.3(b)	
12	(8)Minimum Weld Requirements for Opening in Shells	Figure 5.6	For further details see Clause 5.7.3
13	Shell Manholes	Figure 5.7(a)	
14	Dimensions for Shell Manhole Neck Thickness	Table 5.4(a)	
15	Dimensions for Bolt Circle Diameter and Cover Plate Diameters for Shell Manholes	Table 5.5(a)	
16	Manhole or Nozzle Cross Section	Figure 5.7(a)	
17	Nozzle Cross Section	Figure 5.7(a)	
18	Insert Type Reinforcement for Manholes and Nozzles	Figure 5.7(b)	
19	Reinforcing Plate	Figure 5.8	
20	Regular Type Flanged Nozzles, NPS 80 NB (3″) or Larger	Figure 5.8	

(Continued)

Sl. No.	Description	Figure or Table Reference (API 650)	Remarks
21	Low Type Flanged Nozzles, NPS 80 NB (3″) or Larger	Figure 5.8	
22	Couplings and Flanged Fittings, NPS 20NB ($^3/_4$″) through NPS 50 NB (2″)	Figure 5.8	
23	Shell Nozzle Flanges	Figure 5.10	
24	Dimensions of Shell Nozzles	Table 5.6(a)	
25	Dimensions of Shell Nozzles, Pipes, and Welding Schedules	Table 5.7(a)	
26	Dimensions of Shell Nozzle Flanges	Table 5.8(a)	
27	Flush Type Clean-out Fittings	Figure 5.12	
28	Flush Type Clean-out Fitting Supports	Figure 5.13	Methods A to D
29	Dimensions of Flush Type Clean-out Fittings	Table 5.9(a)	
30	Minimum Thickness of Cover Plates, Bolting Flange, and Bottom Reinforcing Plate for Flush Type Clean-out Fittings	Table 5.10(a)	
31	Thicknesses and Heights of Shell Reinforcing Plates for Flush Type Clean-out Fittings	Table 5.11(a)	
32	Flush Type Shell Connection	Figure 5.14	
33	Dimensions of Flush Type Shell Connection	Table 5.12(a)	
34	Roof Manhole with Reinforcement Plate	Figure 5.16	
35	Base or Roof Manhole without Reinforcement Plate	Figure 5.16	
36	Dimensions of Roof Manholes	Table 5.13(a)	
37	Rectangular Roof Opening with Flanged Covers	Figure 5.17	
38	Rectangular Roof Opening with Hinged Covers	Figure 5.18	
39	Flanged Roof Nozzles	Figure 5.19	
40	Dimensions of Flanged Roof Nozzles	Table 5.14(a)	
41	Threaded Roof Nozzles	Figure 5.20	
42	Dimensions of Threaded Roof Nozzles	Table 5.15(a)	
43	Draw Off Sump	Figure 5.21	
44	Dimensions of Draw Off Sump	Table 5.16(a)	
45	Requirements for Platforms and Walkways	Table 5.17	
46	Requirements for Stairways	Table 5.18	
47	Rise, Run, and Angle Relationship for Stairways	Table 5.19	
48	Scaffold Cable Support	Figure 5.22	
49	Grounding Lug	Figure 5.23	
50	Typical Stiffening Ring Sections for Tank Shells	Figure 5.24	
51	Stairway Opening through Stiffening Ring	Figure 5.25	
52	Some Acceptable Column Base Details	Figure 5.26	

Annexure C: Welding Procedure and Welder Qualification Requirements

Clause 9.2 of API 650 specifies various requirements for welding procedure qualification. Quite often, more stringent requirements would be imposed by clients and consultants based on end users' requirements, especially related to service and constructional features. The restrictions often put forward by clients from the oil and gas industry, especially when dealing with sour hydrocarbon liquid, are provided against relevant clauses of API 650 in the following table.

API 650 Clause	Client Requirements
Welding Procedure Qualification	
9.2.1.1	Welding procedures and qualifications shall be in accordance with Section IX.
	The welding procedure specifications (WPS) for welding pressure retaining and other structural welding shall be as per Section IX.
9.2.1.3	All shell vertical joints shall be welded by shielded metal arc welding (SMAW) with uphill progression or, when specifically approved, automatic electro-gas welding (EGW).
	Horizontal shell joints shall be welded utilizing either SMAW or automatic submerged arc welding (SAW).
	The procedure qualification test for SAW of shell horizontal seams shall be executed in a test frame on plates with a minimum length of 3 m under restrained conditions.
	The procedure qualification for shell vertical seams shall be performed vertically (3G) and shall be subject to impact testing of the weld metal of SMAW welds and weld metal and heat affected zone of EGW welds.
	Qualifications for EGW shall be run on specimens of the full tier height for both the maximum and minimum shell thicknesses. Impact testing of other procedures for thicknesses 38 mm and under shall not be required irrespective of any requirement for impact testing of the plate.
9.2.1.4	Procedure qualification for shell to bottom joints shall be performed on actual joint configuration, material grades, and thicknesses to be used. This procedure qualification test shall demonstrate that the hardness of the heat affected zone at the shell to bottom plate does not exceed 280 HV.
	Hardness testing of procedure qualifications shall be mandatory for Group IV and higher materials. Hardness testing of EGW procedures shall also be mandatory, irrespective of the shell material group. Welds and heat affected zones shall be surveyed. The maximum allowable hardness shall be 280 HV.
	When it is proposed to use a "weldable" protective coating on edges prepared for welding, technical data of coating shall be appended to applicable weld procedures.

(Continued)

API 650 Clause	Client Requirements
	Primers applied principally to protect plate surfaces shall not be considered weldable and shall be locally removed by grinding or brushing prior to welding of attachments, etc.
9.2.2.2	Low hydrogen electrodes shall be used for welding the shell to the bottom joint for all steels with a specified minimum tensile strength in excess of 60,000 psi and for the repair of surface defects.
Welder or Operator Qualification	
9.3.1	All welders for tank erection shall be tested under surveillance of the client or consultant, irrespective of any previous qualification performed.
9.3.2	Welder qualification shall be in accordance with ASME Section IX.
9.3.3	The test may be terminated at the welder's option if the welder becomes aware that he or she has introduced a defect and may be restarted after repreparation or replacement of test specimen once, except for intended restarts for new electrodes or repositioning. On completion of test weld, no repairs shall be allowed even if a defect is revealed by cosmetic grinding or filing.
9.3.3(a)	The test may be terminated by the client or consultant whenever it becomes apparent that the welder lacks the required skill to produce satisfactory results or if he or she fails to observe the requirements of the WPS.
9.3.3(b)	In the event of a test failure, the client or consultant reserves the right to disallow an immediate retest and to insist on a period of further training before any retest.
	Qualification tests using SAW or EGW processes shall in addition to radiographic examination be subject to either ultrasonic examination or bend testing.
	Welders identification cards with photograph, process, base metal, and thickness range qualified shall be issued to all qualified welders under the client or consultant stamp.

Annexure D: Radiography of Storage Tanks

As per Clause 8.1.5 of API 650, radiographs are to be judged according to Clause UW-51(b) of ASME in Section VIII Div (1), "Code for Unfired Pressure Vessels."

Technical Requirements from UW-51 Radiographic Examination of Welded Joints

Indications shown on radiographs of welds and characterized as imperfections are unacceptable under the following conditions and shall be repaired as provided in UW-38, and the repair radiographed to UW-51. (Table reproduced from ASME Section VIII Div (1) with permission.)

1	Any indication characterized as a crack or zone of incomplete fusion or penetration	
2	Any other elongated indication on the radiograph that has length greater than	
3	6 mm (1/4")	for *t* up to 3⁄4" (19 mm)
4	1⁄3 *t*	for *t* from 3⁄4" (19 mm) to 2¼" (57 mm)
5	19 mm (3⁄4")	for *t* over 2¼" (57 mm)

Where *t* = the thickness of the weld excluding any allowable reinforcement. For a butt weld joining two members having different thicknesses at the weld, *t* is the thinner of these two thicknesses.

6 Any group of aligned indications that have an aggregate length greater than *t* in a length of 12 *t*

7 Except when the distance between the successive imperfections exceeds 6 *L* where *L* is the length of the longest imperfection in the group

Rounded indications in excess of that specified by the acceptance standards given in Appendix 4

Terminology as per Appendix 4 of ASME Section VIII Div (1)

(Table reproduced from ASME Section VIII Div (1) Appendix 4 with permission.)

Rounded Indications

Indications with a maximum length of 3× the width or less on the radiograph are defined as rounded indications.

Aligned Indications

A sequence of four or more rounded indications when they touch a line parallel to the length of the weld drawn through the center of two outer rounded indications.

Acceptance Criteria as per Appendix 4 of ASME Section VIII Div (1)

(Table reproduced from ASME Section VIII Div (1) Appendix 4 with permission.)

Image Density

Density within the image of the indication may vary and is not a criterion for acceptance or rejection.

Relevant Indications

Rounded indications that exceed the following dimensions:

1⁄10 *t*	for *t* less than 3 mm (1⁄8″)
0.4 mm (1⁄64″)	for *t* from 3 mm to 6 mm (1⁄8″ to 1⁄4″), inclusive
0.8 mm (1⁄32″)	for *t* greater than 6 mm to 50 mm (1⁄4″ to 2″), inclusive
1.6 mm (1⁄16″)	for *t* greater than 50 mm (2″)

Maximum Size of Rounded Indications

Isolated random	¼ *t* or 4 mm (5/32″), whichever is less
When separated by 25 mm (1″) or more	1/3 *t* or 6 mm (1/4″), whichever is less
For *t* greater than 50 mm (2″)	10 mm (3⁄8″)

Aligned Rounded Indications **(Reproduced from ASME Section VIII DIV (1) Appendix 4 with permission.)**

Acceptable when summation of diameters of indications is less than *t* in a length of 12 *t*.

Aligned rounded indications

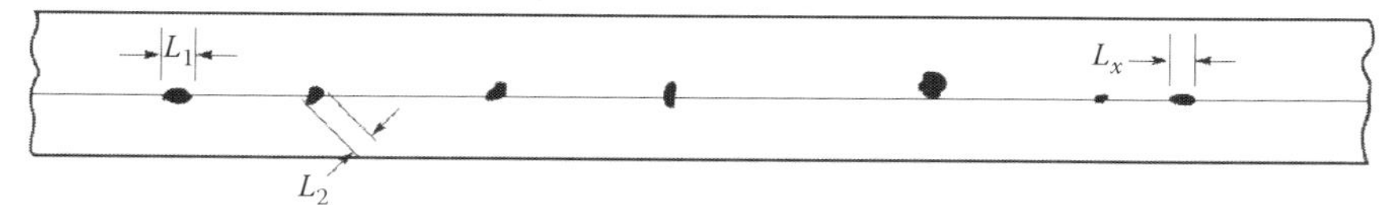

Groups of aligned rounded indications

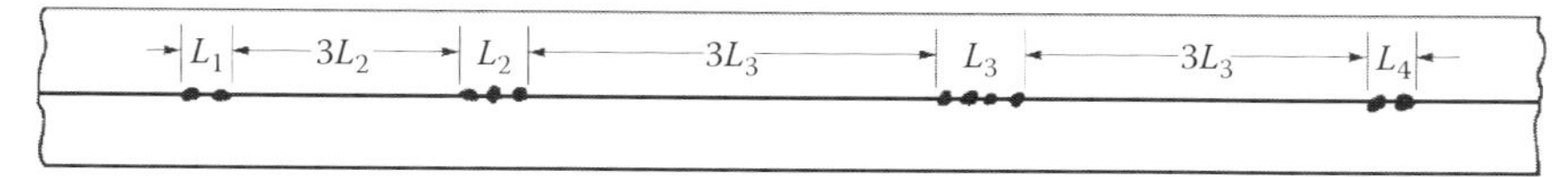

GENERAL NOTE: Sum of the group lengths shall be less than *t* in a length of 12t.

Maximum Group Length	Minimum Group Spacing
L = 1/4 in. (6 mm) for *t* less than 3/4 in. (19 mm)	3*L* where *L* is the length of the longest adjacent group being evaluated.
L = 1/3t for *t* 3/4 in. (19 mm) to 21/4 in. (57 mm)	
L = 3/4 in. (19 mm) for *t* greater than 21/4 in. (57 mm)	

Rounded Indications Charts **(Reproduced from ASME Section VIII DIV (1) Appendix 4 with permission.)**

The rounded indications characterized as imperfections shall not exceed those shown in the following charts.

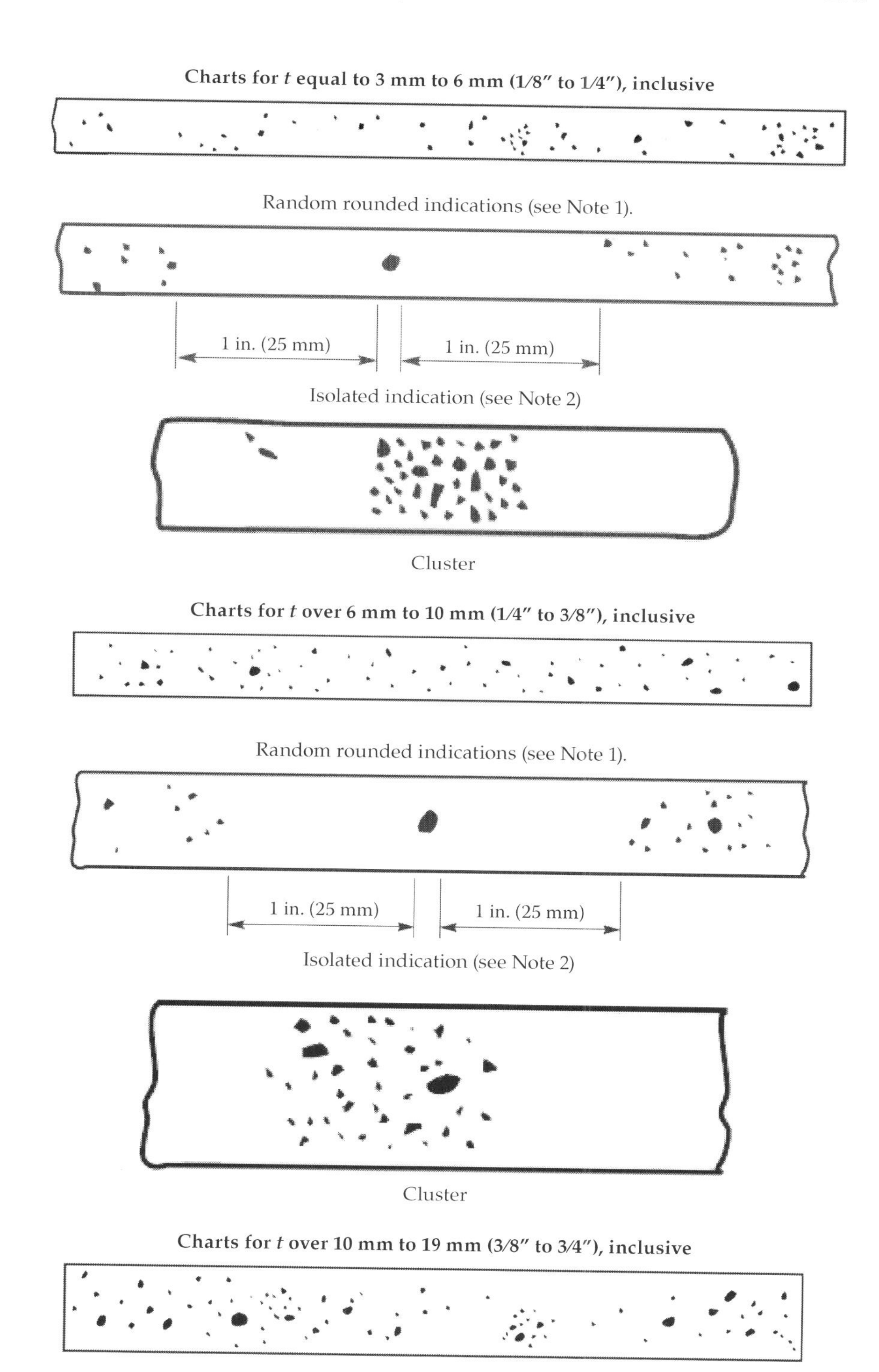
Charts for t equal to 3 mm to 6 mm (1/8" to 1/4"), inclusive
Random rounded indications (see Note 1).
1 in. (25 mm)
1 in. (25 mm)
Isolated indication (see Note 2)
Cluster
Charts for t over 6 mm to 10 mm (1/4" to 3/8"), inclusive
Random rounded indications (see Note 1).
1 in. (25 mm)
1 in. (25 mm)
Isolated indication (see Note 2)
Cluster
Charts for t over 10 mm to 19 mm (3/8" to 3/4"), inclusive

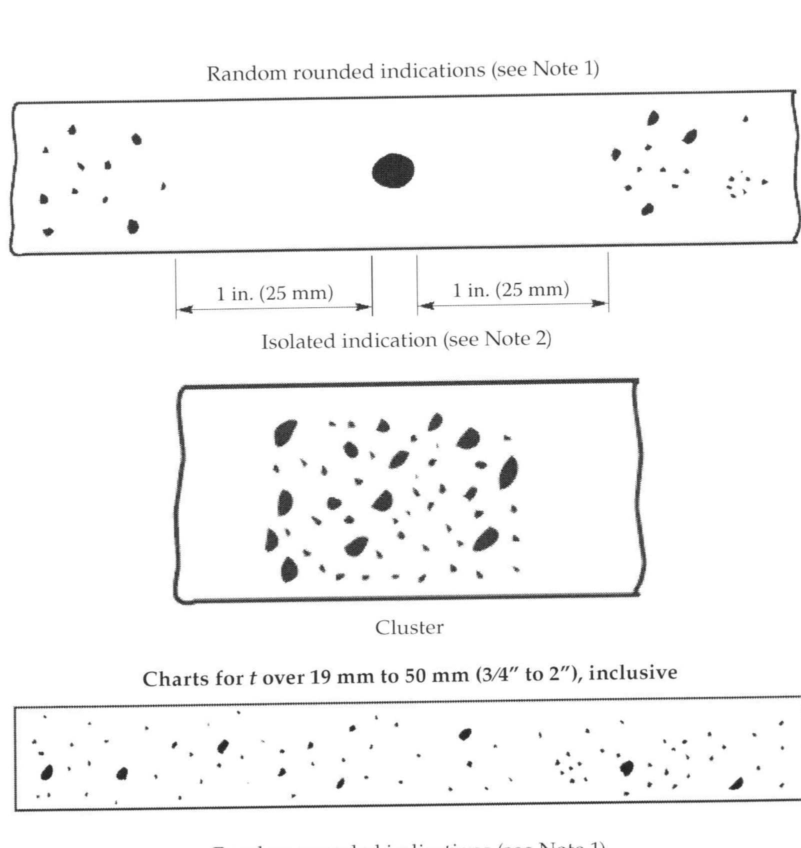

Cluster

Charts for *t* over 19 mm to 50 mm (3/4″ to 2″), inclusive

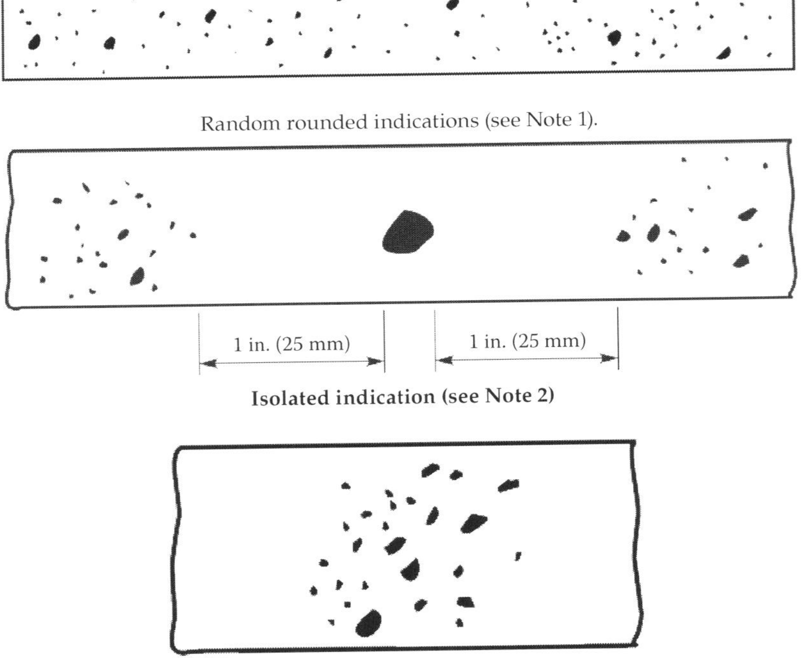

Cluster

Charts for *t* over 50 mm to 100 mm (4″ to 8″), inclusive.

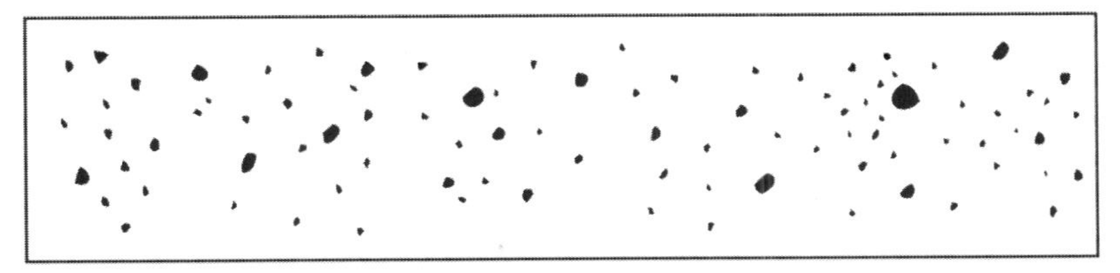

Random rounded indications (see Note 1).

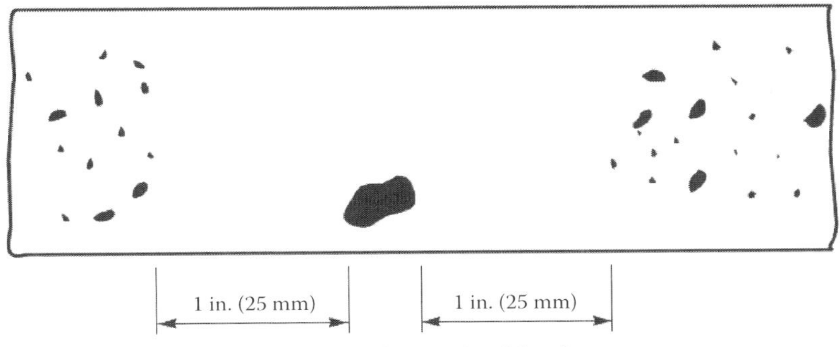

Isolated indication (see Note 2)

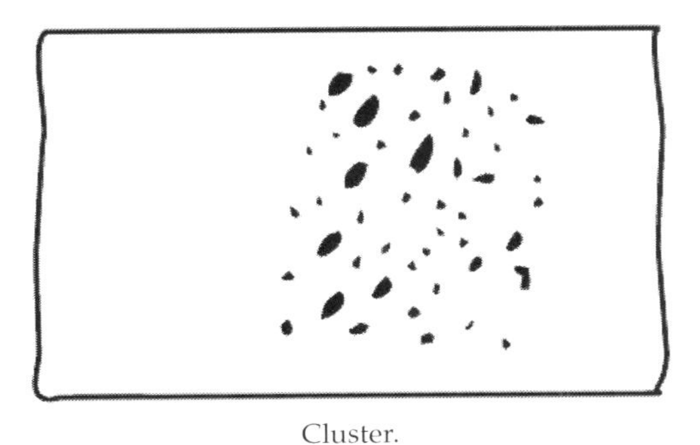

Cluster.

Charts for *t* over 100 mm (4″)

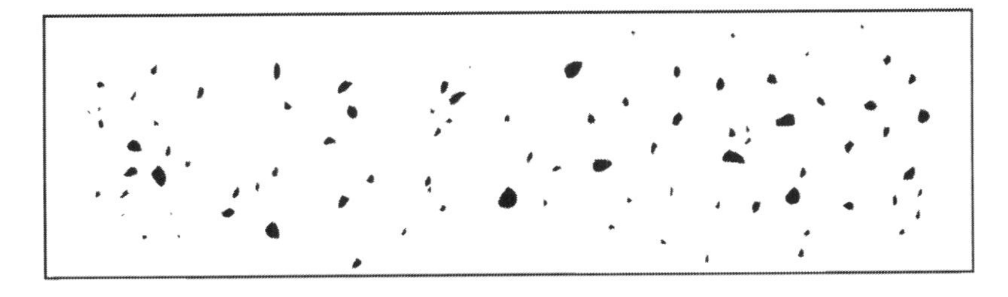

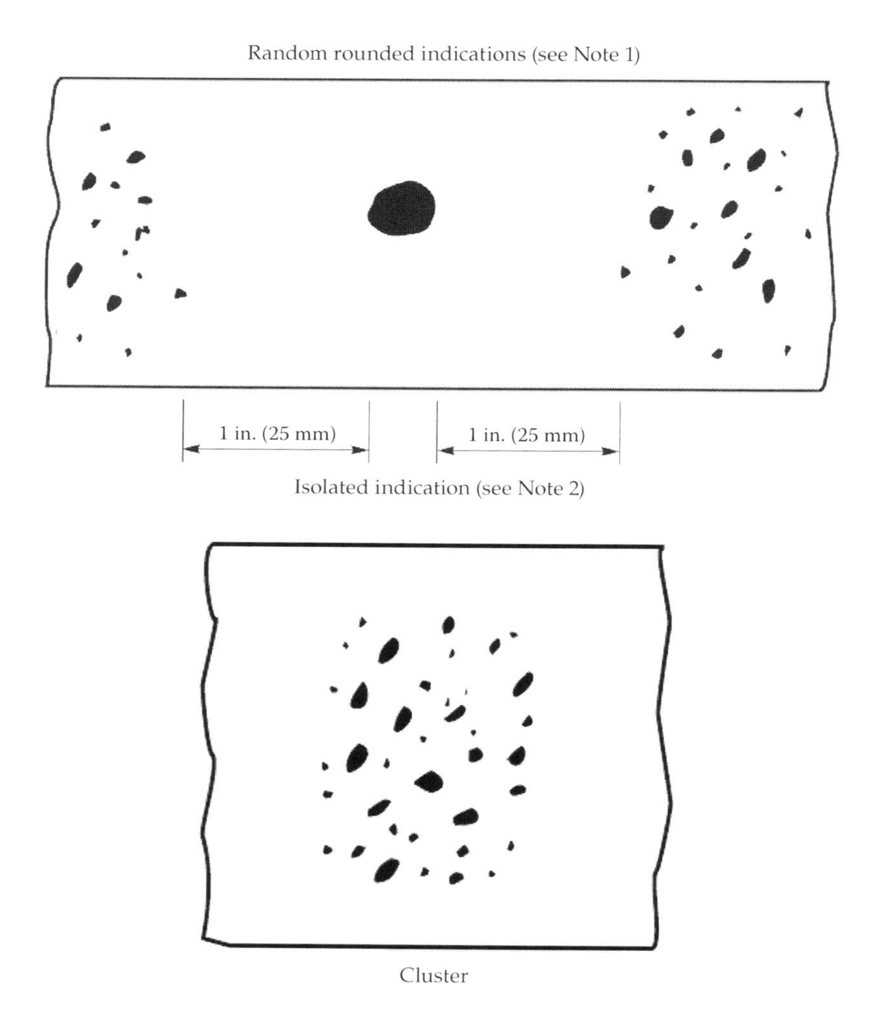

Note (1) Typical concentration and size permitted in any 6 in. (150 mm) length of weld.
Note (2) Maximum size per Table 4-1.

Determination of Limits for Defective Welding

While the excerpts (figures and tables) reproduced from ASME Section VIII Div (1) provide guidelines for estimating the acceptance of each of the defects found in radiographs, Clause 8.1.6 of API 650 provides the guidelines for arriving at the extent of radiography also required for the penal spots to be taken in the case of defective welds. When a radiographed spot shows unacceptable defects, two penalty spots adjacent to the spot radiographed need to be taken. This would be required even when defects extend beyond the boundaries of original spot, until the end of such defective welding. This may extend on one or both sides of the spot radiographed. If it extends to only one side, the penalty need be taken only on one side.

Repair of Defective Welds

Clause 8.1.7 of API 650 requires unacceptable defects to be repaired. As a corollary, these spots need to be radiographed again. The repair of the defective weld shall be carried out by gouging, grinding, or both and rewelded prior to re-radiography.

Annexure E: Inspection and Test Plan

General

The inspection and test plan (ITP) for a storage tank is quite a lengthy document, since it involves the assembly of so many small components, attachments, and fixtures apart from principal components such as the bottom plate, shell, roof, manways, nozzles, and so on. To not miss any such salient activities during inspection, as required in codes and specifications, it is always better to have an ITP covering all these activities so that every activity shall be inspected and recorded.

Though these terms are the general guide words, since tank construction involves multidisciplinary works, it is a usual practice to split the ITP into a few segments as follows, primarily based on the nature and discipline of the work concerned. Depending on the works involved, stage-wise inspections as listed in the following six ITPs would commonly suffice:

1. ITP for foundation till hand over for mechanical erection
2. ITP for mechanical erection of tank
3. ITP for surface preparation and painting
4. ITP for cathodic protection (including provision of sacrificial anodes)
5. ITP for electrical and instrumentation works
6. ITP for subsequent civil works such as preparation of dykes, roads, illumination, and so on

Out of these six ITPs, Item 2 relates to works covered by this book, and hence a typical ITP for a fixed cone roof tank is provided in this annexure. Based on the specific need of each case, the ITP needs to be enlarged or tailored based on the type of the tank, the construction method, and the specific requirements put forward by the client. Readers are encouraged to refer to the comments provided under the remarks column and give serious consideration while formulating the ITP for a specific tank.

Though it does not really fall under the jurisdiction of this book, a typical ITP for surface preparation and painting (Item 3) is also included in this book (in Chapter 13) for the sake of completeness. Since works related to the other four ITPs are not within the ambit of this book, they are not included.

Typical Inspection and Test Plan for Mechanical Works of Storage Tank

No.	Inspection and Test Plan	Reference Document	Acceptance Criteria	Verifying Document	Activity By Manufacturer	TPI	Client	Remark
1	Preinspection meeting	Spec.	Spec.	MOM	H	H	H	[a]
Before Manufacturing								
2	Review of mechanical calculation	Data sheets, PO, Spec.	API 650, Spec.	Calculation book	H	H	A	
3	Review of fabrication drawings	DWG, Spec.	API 650, Spec.	DWG	H	H	A	
4	Review of QA/QC documents	DWG, Spec.	API 650, Spec.	DWG	H	H	A	
Materials								
5	Review mill test certificates of plates	DWG, Certificates	ASME SEC II	Certificates	H	W	R	
6	Review mill test certificates of pipes, fittings, flanges, fasteners, gaskets, etc.	DWG, Certificates	ASME SEC II, ASME B 16.5	Certificates	H	W	R	
7	Review mill test certificates of welding consumable material	WPS, Certificates	ASME SEC II	Certificates	H	W	R	
8	Visual and dimensional check of plates, pipes, fittings, flanges, fasteners, gaskets, etc.	DWG, Spec.	ASME SEC II	Report	H	W	R	
During Manufacturing: Marking, Cutting, Beveling, and Grinding								
9	Annular plates	DWG, WPS, Spec.	API 650, Spec.	Report	H	SW	R	[b]
10	Bottom/sketch plates	DWG, WPS, Spec.	API 650, Spec.	Report	H	SW	R	
11	Shell courses	DWG, WPS, Spec.	API 650, Spec.	Report	H	SW	R	

No.	Inspection and Test Plan	Reference Document	Acceptance Criteria	Verifying Document	Activity By			Remark
					Manufacturer	TPI	Client	
12	Roof plates	DWG, WPS, Spec.	API 650, Spec.	Report	H	SW	R	[b]
13	Roof structures	DWG, WPS, Spec.	API 650, Spec.	Report	H	SW	R	[c]
14	Wind griders (if required)	DWG, WPS, Spec.	API 650, Spec.	Report	H	SW	R	[b]
15	Top angles (if required)	DWG, WPS, Spec.	API 650, Spec.	Report	H	SW	R	[c]
16	Reinforcement pads	DWG, WPS, Spec.	API 650, Spec.	Report	H	SW	R	[b]
17	Manhole nozzle neck (from plates)	DWG, WPS, Spec.	API 650, Spec.	Report	H	SW	R	
18	Pipe for column (if required)	DWG, WPS, Spec.	API 650, Spec.	Report	H	SW	R	[d]
19	Pipe for nozzles	DWG, WPS, Spec.	API 650, Spec.	Report	H	SW	R	[d]
20	Internal pipes	DWG, WPS, Spec.	API 650, Spec.	Report	H	SW	R	[d]
21	Internal or external supports	DWG, WPS, Spec.	API 650, Spec.	Report	H	SW	R	[d]
22	Sumps	DWG, WPS, Spec.	API 650, Spec.	Report	H	SW	R	[b]
23	Spiral stairway and platforms	DWG, WPS, Spec.	API 650, Spec.	Report	H	SW	R	
24	Visual and dimensional check of loose parts	DWG, WPS, Spec.	API 650, Spec.	Report	H	H	R	[e]
During Manufacturing: Forming and Rolling								
25	Shell courses	DWG, Spec.	API 650, Spec.	Report	H	W	R	
26	Sumps	DWG, Spec.	API 650, Spec.	Report	H	SW	R	
27	Top angles	DWG, Spec.	API 650, Spec.	Report	H	W	R	
28	Manway nozzle neck (from plates)	DWG, Spec.	API 650, Spec.	Report	H	SW	R	
29	Visual and dimensional check	DWG, Spec.	API 650, Spec.	Report	H	H	R	
During Manufacturing: Foundation								
30	Levelness of ring wall (if applicable)	DWG, Spec.	API 650, Spec.	Report	H	H	R	[f]

(*Continued*)

No.	Inspection and Test Plan	Reference Document	Acceptance Criteria	Verifying Document	Activity By			Remark
					Manufacturer	TPI	Client	
31	Center point and radius	DWG, Spec.	API 650, Spec.	Report	H	H	R	f
32	Flatness and slope	DWG, Spec.	API 650, Spec.	Report	H	H	R	f
33	Orientation	DWG, Spec.	API 650, Spec.	Report	H	H	R	f
34	Anchor bolts (distance, projection, etc.) as required	DWG, Spec.	API 650, Spec.	Report	H	H	R	f
During Manufacturing: Welding								
35	WPS and PQR	DWG, WPS, PQR, Spec.	ASME SEC IX	Report	H	R/A		g
36	Welder/welding operator qualification	DWG, WPS, WQT, Spec.	ASME SEC IX	Report	H	R/A	R	h
37	NDT operator certificate check	Certificates	API 650, Spec.	Report	H	R/A	R	
38	Fit up of bottom plates	DWG, WPS, Spec.	API 650, Spec.	Report	H	H	R	
39	Welding sequence implementation	DWG, WPS, Spec.	API 650, Spec.	Report	H	H	R	
40	Visual inspection of weldment	DWG, WPS, Spec.	API 650, Spec.	Report	H	H	R	i
41	Vacuum box test of bottom plates	Spec.	API 650, Spec.	Report	H	H	R	i
42	Annular plate orientation check	DWG, Spec.	API 650, Spec.	Report	H	H	R	
43	Fit up and welding of annular plate	DWG, WPS, Spec.	API 650, Spec.	Report	H	H	R	
44	NDT of annular plate/vacuum test (if required)	Spec.	API 650, Spec.	Report	H	H	R	
45	Fit up of shell plate	DWG, WPS, Spec.	API 650, Spec.	Report	H	H	R	
46	Welding of shell plate	DWG, WPS, Spec.	API 650, Spec.	Report	H	H	R	
47	PWHT (if required)	WPS, Spec.	API 650, Spec.	Report	H	H	R	j
48	NDT of shell plate	Spec.	API 650, Spec.	Report	H	H	R	
49	Oil test of annular plate to shell plate	Spec.	API 650, Spec.	Report	H	H	R	i

No.	Inspection and Test Plan	Reference Document	Acceptance Criteria	Verifying Document	Activity By Manufacturer	TPI	Client	Remark
50	Fit up and welding compression ring and roof plate	DWG, WPS, Spec.	API 650, Spec.	Report	H	H	R	
51	Welding of roof plates	DWG, WPS, Spec.	API 650, Spec.	Report	H	H	R	
52	Air test of roof plate	Spec.	API 650, Spec.	Report	H	H	R	
53	Air test of reinforcement pads	Spec.	API 650, Spec.	Report	H	H	R	
During Manufacturing: Dimensional Check								
54	Plumbness	DWG, Spec.	API 650, Spec.	Report	H	H	R	
55	Roundness	DWG, Spec.	API 650, Spec.	Report	H	H	R	
56	Peaking and banding	DWG, Spec.	API 650, Spec.	Report	H	H	R	
57	Nozzle elevation, orientation, and projection	DWG, Spec.	API 650, Spec.	Report	H	H	R	
58	Manhole elevation, orientation, and projection	DWG, Spec.	API 650, Spec.	Report	H	H	R	
59	Major overall dimensions	DWG, Spec.	API 650, Spec.	Report	H	H	R	
During Manufacturing: Hydrostatic Test								
60	Verification of test package	Reports	API 650, Spec.	Report	H	H	R	
61	Hydro test (water level/ holding time)	Spec.	API 650, Spec.	Report	H	H	H	
62	Settlement check (refer to applicable procedure)	DWG, Spec.	API 650, Spec.	Report	H	H	R	
63	Water discharge	Spec.	API 650, Spec.	Report	H	W	R	
64	Cleaning check	Spec.	API 650, Spec.	Report	H	H	R	
During Manufacturing: Painting Activities								
65	Painting material identification/ check	Spec.	API 650, Spec.	Report	H	H	R	

(Continued)

No.	Inspection and Test Plan	Reference Document	Acceptance Criteria	Verifying Document	Activity By			Remark
					Manufacturer	TPI	Client	
66	Weather condition and surface preparation (blasting)	Spec.	API 650, Spec.	Report	H	H	R	k
67	Primer layer inspection (including dry thickness check)	Spec.	API 650, Spec.	Report	H	H	R	k
68	Intermediate/ final layer inspection (including dry thickness check)	Spec.	API 650, Spec.	Report	H	H	R	k
During Manufacturing: Cathodic Protection and Electrical								
69	Check of material prior to installation	Spec.	API 650, Spec.	Report	H	H	R	
70	Installation of earthing bosses	Spec.	API 650, Spec.	Report	H	H	R	
71	Anode installation	Spec.	API 650, Spec.	Report	H	H	R	
72	Check of anode and conductor	Spec.	API 650, Spec.	Report	H	H	R	
73	Check the bonding of UG pipe	Spec.	API 650, Spec.	Report	H	H	R	
74	Cable installation	Spec.	API 650, Spec.	Report	H	H	R	
75	Check the test box, reference electrode	Spec.	API 650, Spec.	Report	H	H	R	
76	Installation of reference electrode including cable	Spec.	API 650, Spec.	Report	H	H	R	

No.	Inspection and Test Plan	Reference Document	Acceptance Criteria	Verifying Document	Activity By			Remark
					Manufacturer	TPI	Client	

Note: TPI = third-party inspection. H = hold point: hold on the production till the TPI inspector performs the inspection and supervises the required test. W = witness point: the manufacturer shall notify the client and TPI inspector, but there is no hold on the production. The client can waive this inspection based on his or her discretion and informs the TPI inspector accordingly. R = review document, which includes material test certificates, WPS, PQR, NDT procedures, etc. A = approval. SW = spot witness: for items with spot witness, manufacturer shall notify the TPI inspector as fulfilling the monitoring; for example, one random visit for whole UT tests or one or two visits for whole surface preparation works for painting. MOM = minutes of meeting. PO = purchase order. 1. As far as possible respective specification numbers and procedure numbers shall be included in the ITP. 2. Provide applicable format numbers for various inspection reports. 3. Intervention levels proposed are quite arbitrary and shall vary according to the expertise of the contractor and the confidence of the client. 4. Intervention by all agencies at every stage may cause delay because of coordination issues and hence should be decided judiciously. 5. In addition to the notified inspection stages, it is always better to have undeclared surveillance visits to assess the true quality of work. 6. Each activity shall be signed off by all the parties involved, and the date of signing shall indicate the completion date to avoid ambiguity.

[a] Review document status, GA, detailed drawings, QA/QC documents such as WPS, PQR, WQT, NDT, and other test procedures.

[b] Each plate or component identified in the cutting plan to covered by a report.

[c] Each load-bearing structural member to be covered by an inspection report (against part numbers).

[d] Each pipe length shall be identified against part numbers in the drawing for which it is used.

[e] Match marking to be done wherever possible.

[f] Civil engineering activity. Verification of reports alone in envisaged, as a cross verification is always recommended.

[g] Already established WPS or PQR also could be considered.

[h] Preferably, all welders are to be qualified at the site.

[i] One hundred percent check (positively) before the hydrostatic test.

[j] Verify chart speed, heating and cooling rates, and soaking time and temperature.

[k] Only an overview. Detailed one provided in Chapter 13.

Annexure F: Requirements for Floating Roof Tanks

General

Tanks with floating roofs on a tank widely used in the oil industry for decreasing vapor loss when volatile oil products are to be stored in the tank. Floating roofs can be used when the tanks are open at the top or when provided with a fixed roof. They can also be used when tanks are closed and no product leak is permitted to the environment.

Since the most extensively used type of floating roof tank (in oil and gas) is the external floating roof type, this annexure elaborates on special requirements for this type of tank. However, for the sake completeness, the internal floating roof type is also touched upon briefly.

The functioning of floating roofs is strongly dependent on the good geometry of the shell, floating roof, and antirotation devices and hence requires careful considerations during construction and maintenance. The main advantages of tanks with floating roofs are as follows:

1. Minimizes loss of the stored product through evaporation. Factors influencing product evaporation are as follows:
   a. Temperature of liquid
   b. Presence and size of vapor space above liquid
   c. Possibility for ventilation of vapor space
   d. Presence of free surface of stored product
2. Decreased evaporation of the product reduces corrosion of the upper parts of the shell.
3. It lowers the risk of fire because of reduced evaporation.
4. The roof load is not transmitted to the shell.

The factors that influence tank construction and type of floating roof are as follows:

1. Characteristic of stored product
2. Climatic conditions at the site
3. Volume of stored product
4. Operational flexibility

Types of Floating Roof Tanks

A typical external floating roof tank consists of an open-topped cylindrical steel shell equipped with a roof that floats on the surface of the stored liquid, rising and falling with the liquid level. The floating roof is composed of a deck, fittings, and rim seal system. Floating roof decks are constructed of welded steel plates and are of three general types: pan, pontoon, and double deck.

Although numerous pan-type decks are currently in use, the present trend is toward pontoon and double-deck-type floating roofs. The two most common types of external floating roof tanks are shown in the following sketches.

Manufacturers supply various versions of these basic types of floating decks, which are tailored to emphasize particular features, such as full liquid contact, load-carrying capacity, roof stability, and pontoon arrangement. The liquid surface is covered by a floating deck, except in the small annular space between the deck and the shell. The deck may contact liquid or float directly above the surface on pontoons.

External Floating Roof Tank

External floating roof tanks are equipped with a rim seal system, which is attached to the roof perimeter and contacts the tank wall. The rim seal system slides against the tank wall as the roof is raised and lowered. The floating deck is also equipped with fittings that penetrate the deck and serve operational functions. The external floating roof design is such that evaporative losses from stored liquid are limited to losses from the rim seal system and deck fittings (standing storage loss) and any exposed liquid on the tank walls (withdrawal loss). In the external floating roof design, the roof is made to rest on the stored liquid and is free to move with the level of the liquid. These tanks reduce evaporation losses and control breathing losses while filling. They are preferred for storage of petroleum products with a true vapor pressure of 10.3 kPa to 76.5 kPa absolute.

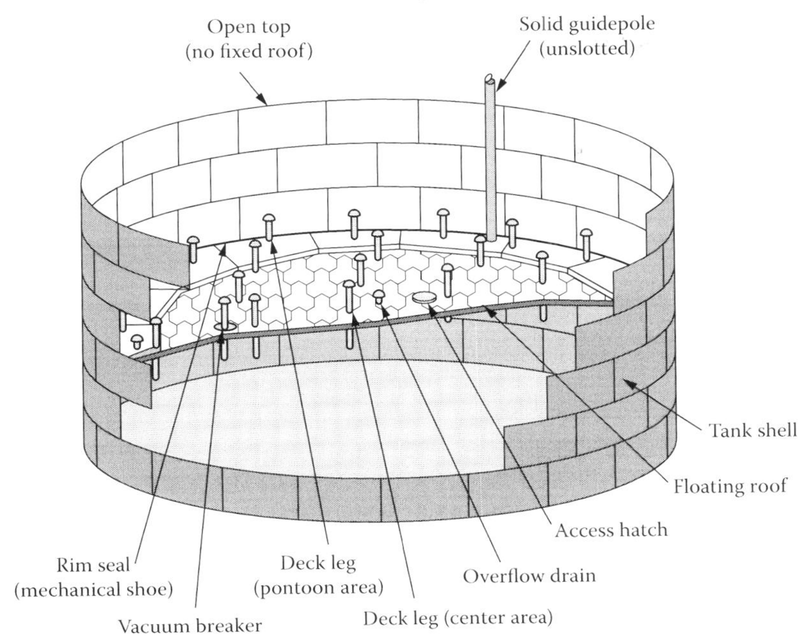
External floating roof tanks
Open top
(no fixed roof)
Solid guidepole
(unslotted)
Tank shell
Floating roof
Access hatch
Rim seal
(mechanical shoe)
Deck leg
(pontoon area)
Overflow drain
Vacuum breaker
Deck leg (center area)

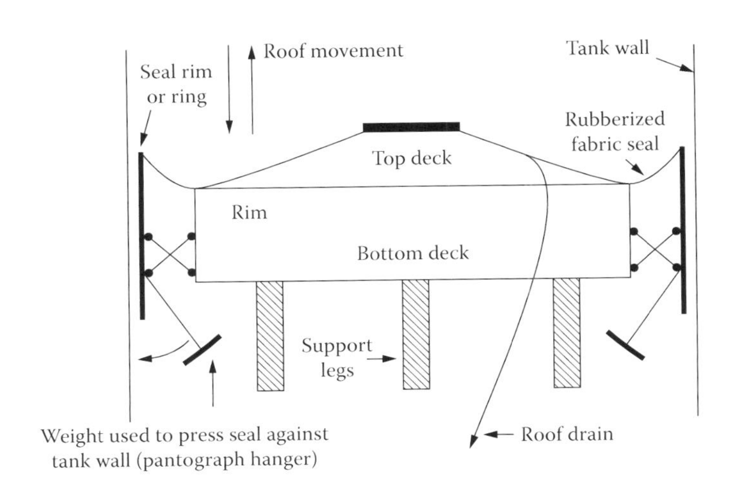
Roof movement
Tank wall
Seal rim
or ring
Rubberized
fabric seal
Top deck
Rim
Bottom deck
Support
legs
Weight used to press seal against
tank wall (pantograph hanger)
Roof drain

Typical List of Floating Roof Fittings

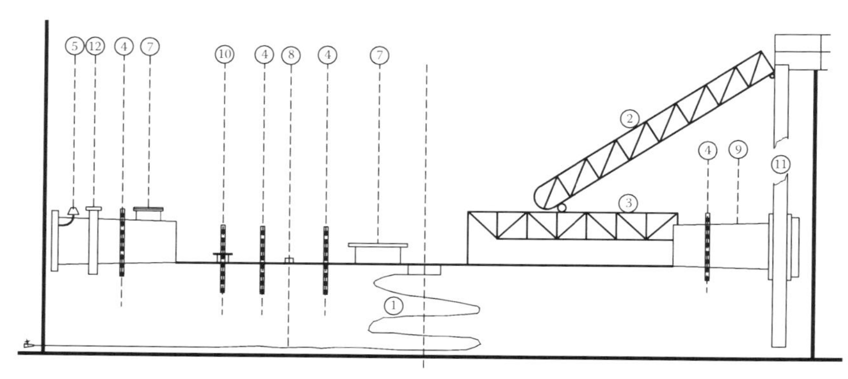

Elevation

1. Roof Drain
2. Rolling Ladder
3. Ladder Runway
4. Support Legs
5. Rim vent
6. Deck Manhole
7. Pontoon Manhole
8. Drain Plug
9. Pontoon
10. Auto Bleeder Vent
11. Roof Guide Pole and Manual Dipping Tube
12. Sample Hatch

Internal Floating Roof Tank

An internal floating roof tank has both a permanent fixed roof and a floating roof inside. There are two basic types of internal floating roof tanks: tanks in which the fixed roof is supported by vertical columns within the tank, and tanks with a self-supporting fixed roof and no internal support columns. A fixed roof is not necessarily free of openings but does span the entire open plan area of the vessel. Fixed roof tanks that have been retrofitted to employ an internal floating roof are typically of the first type, while external floating roof tanks that have been converted to an internal floating roof tank typically have a self-supporting roof.

Primary Subclassification of Floating Roofs

Double-Deck Floating Roof

The entire roof is constructed of closed-top flotation compartments.

Single-Deck Pontoon Floating Roof

The outer periphery of the roof consists of closed-top pontoon compartments, with the inner section of the roof constructed of a single deck without flotation means.

Standard and Client Requirements for Floating Roof Tanks

Annex C of API 650 regarding external floating roofs provides additional requirements applicable to a floating deck. As in the case of the main tank (bottom plate, shell, and roof), here also additional requirements are stipulated by clients based on their specific requirements, especially for the sour environment in the oil and gas industry. Salient requirements specified by clients in the oil and gas industry with a reference to the applicable clauses in API 650 are shown in the following table as a quick reference. As this is provided as a guideline, for finer details and other applicable conditions associated with these requirements, refer to the API Standard.

Sl. No.	API Code Clauses	Client Requirements from Oil and Gas
Scope		
1	C.1.1	Pan-type roofs are not permitted.
Materials		
2		As per data sheet.
Design		
3		No additional requirements.
Decks		
4	C.3.3.2	Minimum thickness of bottom deck shall be 6 mm.
5	C.3.3.3	All compartments shall be internally seal welded along the top, bottom, and vertical edges. Top surfaces, including outer rim, shall be painted.
6	C.3.3.4	For a double-deck roof, the bottom deck is to slope upward toward the center of the tank. The top deck may either slope to drain toward the center (converging double-deck type) or slope down from both the rim and the center.
Pontoon Design		
7		No additional requirements.
Pontoon Opening		
8	C.3.5	Compartments shall be inspectable.

(*Continued*)

Sl. No.	API Code Clauses	Client Requirements from Oil and Gas
Compartments		
9	C.3.6	Compartments shall be inspectable.
Ladders		
10	C.3.7	A rolling ladder with rails is required for all floating roof tanks. Rails shall be welded to pads and not directly onto the roof deck. The ladder shall connect to the gauger's platform. Pivot and wheel bearings shall be of a type that does not require any lubrication after commissioning. Wheels shall be nonsparking. The slope of the ladder shall not exceed 1270 mm (50″) to horizontal at the extreme low position. Treads shall be self-leveling. The minimum inside width of the ladder shall be 750 mm (30″). If it is necessary for rails to be significantly elevated, then suitable steps shall be provided down to the deck. It should be designed for wind and rain conditions prevailing at the site.
11		A 35 mm^2 flexible copper conductor shall be applied across ladder hinges, between the ladder and the tank top, and between the ladder and the floating roof.
Roof Drains		
Primary		
12	C.3.8.1(1)	The drain shall be valved at the tank shell.
13		For the pontoon type, the check valve shall be accessible from the deck by mounting it in a recess.
14		The maximum nominal size of articulated drain lines is 100 mm (4″). If insufficient, additional numbers are to be provided.
15	C.3.8.1(3)	Primary roof drains shall be either articulated joint or flexible hose type.
Emergency Drain		
16	C.3.8.2	Inside the tank, the drain shall be fitted with a deflector to minimize vapor losses.
Out-of-Service Supplementary Drains		
17	C.3.8.3	No additional requirements.
Maintenance Drain		
18		Pontoon-type roofs shall be provided with two numbers of NPS 80 (3″) brass drain plugs, one located close to the center and the other close to the inner rim and fitted from the top for use when the roof is landed on its supports. Drainage of double-deck roofs may be achieved by removing (unscrewing) the emergency drain.
Vents		
19	C.3.9	Design, size, and location of automatic bleeder vents shall be adequate to breathe air and vapors at a volumetric rate at least equal to the maximum tank outflow and to twice the specified commissioning liquid inflow. At least one bleeder vent shall be located near the center of the deck.

Sl. No.	API Code Clauses	Client Requirements from Oil and Gas
20		Pontoon-type roofs shall also be provided with two weighted pressure relief valves to discharge excess tank vapor and incondensibles while the roof is afloat. One of these devices shall be located within 1 m or 2 m from the center, and the other shall be located adjacent to the inner rim and oriented 180° from the pivot of a rolling ladder. The set pressure of each valve shall be 25% above the pressure exerted by the weight of the center deck and its fittings so that there will be no discharge under normal operation but only in the event of a pronounced "ballooning" of the center deck.
21		A minimum of one rim vent shall be installed for each tank.
Supporting Legs		
22	C.3.10.1	Supporting legs shall be adjustable so that whatever the grounding height for operation is, the height can be set to give 2 m clearance up to the lowest deck for maintenance.
23 24		Drain holes at the bottom of support legs shall be ample to ensure that blockage by sludge does not prevent legs from draining completely when the tank is emptied.
25	C.3.10.2	Roof supports shall be designed to support the weight of the roof with two adjacent flooded compartments and with 1.2 kPa (25 lb/ft^2) uniform loading of the center deck of pontoon-type roofs.
26	C.3.10.3(b) C.3.10.4 C.3.10.5 C.3.10.6	Pads of minimum 600 mm diameter shall be provided to distribute support loads at the landing of legs to the tank bottom. Pads shall be continuously welded to the tank bottom. After lining this pad along with the tank bottom, an additional pad shall be provided to avoid damage to the bottom lining by legs. This second pad shall be slightly smaller than the first and shall be located and lightly secured by an epoxy fillet around its edges.
27	C.3.10.7	Sleeves shall be adequately reinforced by means of topside pads and gussets but shall also be welded to the underside of the deck.
28	C.3.10.8	Upward projection of support leg sleeves through the center deck shall be sufficient to ensure that the product cannot flow onto the deck under either of the conditions described in C.3.4.1 of API 650.
Roof Manholes		
29	C.3.11	Two 750 mm (30") deck hatches are to located diametrically opposite to provide access from the underside to the topside of the roof during maintenance.
30		A second manway shall be provided for each double-deck buoyancy compartment.
Centering and Antirotation Devices		
31	C.3.12.2	Roller-type devices shall be provided with grease nipples.
Peripheral Seals		
32	C.3.13.1	The roof perimeter shall incorporate a flexible double sealing mechanism.

(Continued)

Sl. No.	API Code Clauses	Client Requirements from Oil and Gas
33	C.3.13.2	Details of seal material required shall include data on the following:
	C.3.13.3	a. Suitability and corrosion resistance of nonmetallic materials against products stored
	C.3.13.4	b. Suitability up to 80°C
	C.3.13.5	c. Fire-retardant properties of nonmetallic materials
	C.3.13.5(a)	d. Foam application considerations
	C.3.13.5(b)	e. Rim gap tolerances necessary to maintain the seal
		f. Operational constraints for inspection, maintenance, and component replacement
		g. Whether maintenance can be performed on secondary sealing without affecting primary sealing
		h. Self-cleaning characteristics
		i. Means of electrical insulation of mechanical shoe hangers
		j. Drainage of seal system for maintenance
Mechanical Shoe Seal		
34	C.3.13.5(c)	Irrespective of the type of primary or secondary rim sealing, a weather shield shall also be provided, especially for protection from dust.
Gauging Devices		
35	C.3.14.1 C.3.14.1(1) C.3.14.1(2)	Two slotted still wells of minimum 200 mm (8") diameter shall be provided for manual dipping and for level gauging. These shall be located close to the shell so that they are operable from the gauger's platform or wind girder access way.
36	C.3.14.2 C.3.14.3	Still wells for manual dipping shall be provided with a hinged brass cover to preclude the risk of sparking. A semicircular top reference plate, notched 25 mm wide × 40 mm deep, shall be welded inside the manual still well.
37	C.3.14.4	The lower end of still well pipes shall be restrained from lateral movement by sleeves supported from the shell.
38	C.3.14.5	A datum plate of bare steel shall be provided below the manual still well, mounted level on a lined bottom, and held in place by resin.
Inlet Diffuser		
39	C.3.1.5	No additional requirements.
Other Roof Accessories		
40	C.3.16.3	A sampling hatch with a hinged cover shall be brass contact to preclude the risk of sparking. A bare steel datum plate shall be provided below the sampling hatch to enable this fitting to be used as an alternative gauging device. This shall be mounted level on a lined bottom and held in place by resin, as for the manual still well.
Automatic Level Gauge		
41	C.3.16.4	No additional requirements.

Sl. No.	API Code Clauses	Client Requirements from Oil and Gas
Side Entry Mixers		
42	C.3.16.5	No additional requirement.
Fabrication, Erection, Welding, Inspection, and Testing		
43	C.4	No additional requirements.

Typical Floating Roof Designs

This annexure deals only with external floating roofs, being the most commonly used in the oil and gas industry. Appendix C of API 650 defines various requirements and considerations for the design of such roofs, including materials, minimum thickness, and other design conditions to be fulfilled. Various types of floating roof designs used in the industry are provided next.

Out of the five types of roof designs explained next, the most common and predominant ones are the types indicated in Items 2 and 4.

1. Single Deck, without Pontoons and with Supporting Truss on Membrane

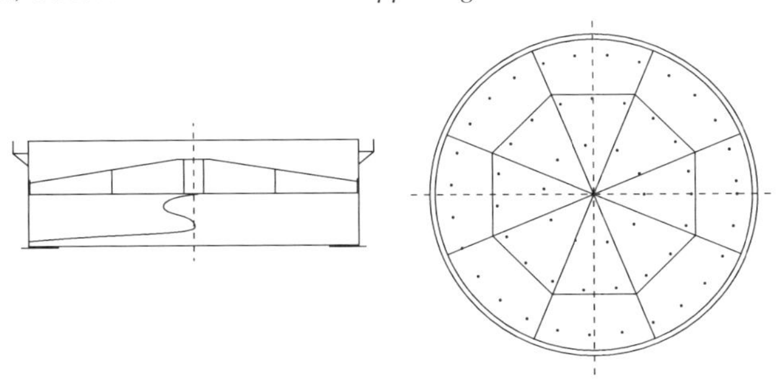

One of the first kinds of floating roofs, it is installed on tanks with a diameter up to 45 m. Because of its high boards, it can float even when there is water or snow on it.

Advantages	Disadvantages
Simplicity in construction and maintenance	Has no positive buoyancy and sinks if membrane is punctured
	Wind can push accumulated water on the roof toward one board, which may lead to bending and eventual sinking

2. Single Deck, with Outsider Ring of Pontoons on Periphery and Membrane in Middle

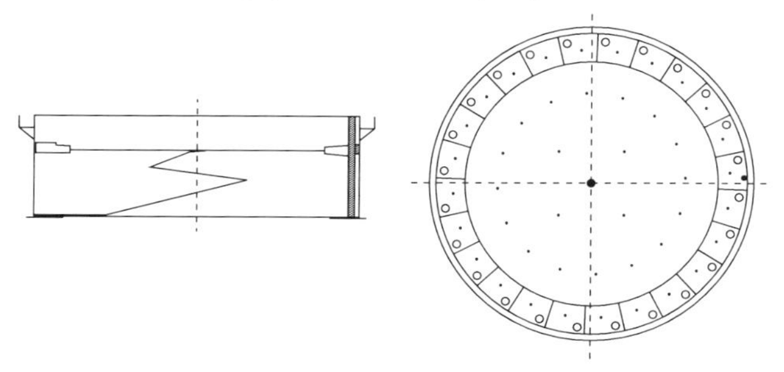

This type is used in regions with low wind pressure so it is not possible to tear the membrane from the product and in regions with little sunshine where the lower temperature of the membrane reduces evaporation of the product. It is often recommended for tanks with a diameter ≤ 50 m.

Advantages	Disadvantages
Simple in construction and can be prefabricated and mounted easily	Increased possibility of membrane deformation
Less costly per unit of covered surface	Undulations on membrane work as an obstacle for water drain off
No special equipment needed for mounting	Difficult to maintain slope on membrane for proper drain off

3. Single-Deck Floating Roof with Peripheral Ring of Pontoons and One Central Pontoon

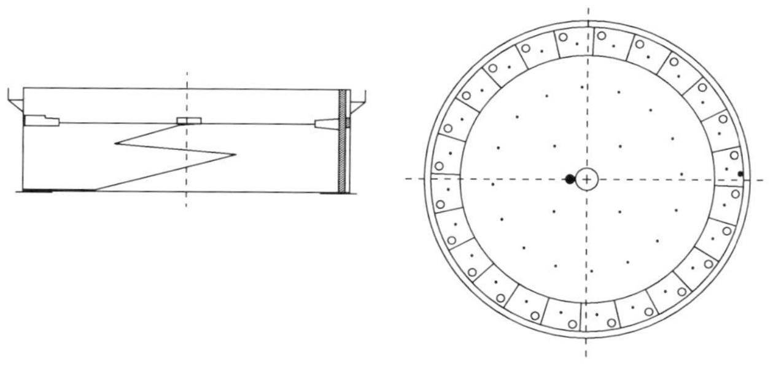

The type is used in regions where wind pressure is more, which may cause tearing of the membrane from the product. It is often recommended for tanks with a diameter > 50 m.

Advantages	Disadvantages
Improved geometrical form of the central part	Complicated roof drain system
Increased buoyancy of the roof even when the membrane is punctured	
Simple in construction and for maintenance	
Relatively less cost per unit of covered surface	

4. Double-Deck Floating Roof, with Two Membranes Covering All Surfaces of Roof and Structure

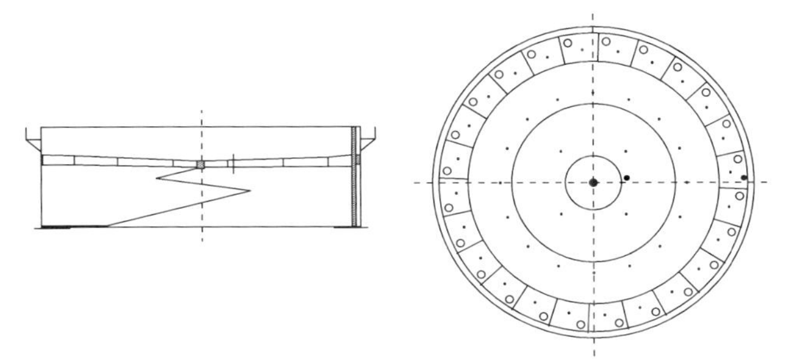

This type is used in regions with more wind pressure, which may cause tearing of the membrane from the product, and in regions with more sunshine where it is necessary to reduce the temperature of the membrane in contact with the product. It is often recommended for tanks with a diameter > 50 m.

Advantages	Disadvantages
Construction has better stiffness, and geometrical form is well supported	Increased cost (metal and labor) per unit surface covered
Increased buoyancy of the roof compared with that of a single-deck roof	More welding in a confined space
Lower deformation when the snow load is irregular because of wind blast	Many erection devices needed
Better stability of the roof during wind blasts	

5. Single-Deck Floating Roof with Ring of Pontoons in Periphery and Many More on Membrane

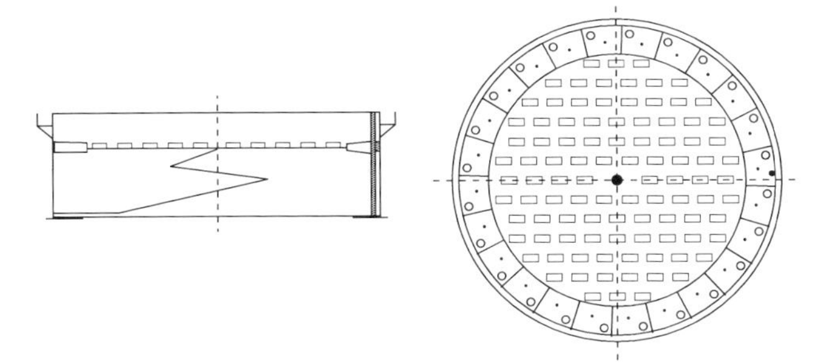

This type combines the strengths of single-deck and double-deck floating roofs. It is used in regions with increased wind pressure, which may cause tearing of the membrane from the product, and is often recommended for tanks with a diameter > 50 m.

Advantages	Disadvantages
Simple in construction	Increased cost (metal and labor) per unit of surface
Does not require special erection devices	Increased length of weld
Increased buoyancy compared to that of a single-deck roof	
Lower deformation when snow load is uneven because of wind blast	
More stable even in wind blasts	

Pictures of Floating Roof Tank Accessories

Rolling ladder and antirotation device.

Pontoon construction.

Floating roof support leg and sleeve (high leg position).

Reinforcing pad for roof leg.

Annexure G: Additional Inspections and Tests for Floating Roof Tanks

Additional Tolerance Required in Construction

For tanks to have an acceptable appearance and to permit proper functioning of a floating roof, they shall have strict tolerances as specified in API 650 or the client specification.

Tolerance allowed on the annular clearance between the shell and the floating roof shall be compatible with the requirement specified by the seal manufacturer but shall be limited to 50 mm radially.

Procedure for Rainwater Accumulation and Puncture/Floatation Tests

The rainwater accumulation test is to ascertain smooth operation of the floating deck even when roof drains are blocked. The puncture/floatation test is carried out to doubly make sure the floating roof operates even when a few pontoons or compartments in the floating deck are punctured during service. Such a test would facilitate the operating plant to get some breathing time to plan for a shutdown when it starts noticing a tilt of the floating deck, resulting from a puncture in a pontoon or floatation compartment.

Checklist Prior to Tests on Floating Roofs

1. Ensure completion of erection and visual inspection (100%) nondestructive testing (as specified).
2. This is followed by tests such as pneumatic testing, diesel chalk testing of the outer rim, and so on. In case any rectifications or repairs

are needed, they shall be completed and the roof reinspected to a satisfactory level.

3. Non-return valve (NRV) in central roof drain shall be checked for its functioning.
4. Ensure that tests such as the diesel chalk test, liquid penetrant testing (LPT), reinforcement pad (RF.Pad) pneumatic test, vacuum box test, and so on are also completed.
5. Punctures with couplings fitted in the deck and adjacent pontoons to be ensured before commencing of hydrostatic testing.
6. The puncture test shall be carried out when the roof is in the highest position.

Test Medium

The test medium shall be iron-free water, which is clean and free of silt. Potable water is a good option.

Procedure for Puncture Test

1. Initial reading of rim space (at least at eight points) to be taken.
2. Location of rolling ladder (drum) with respect to the shell and spirit level reading of threads of staircase need to be taken.
3. Deck immersion and rim position with respect to the tank shell shall also be recorded.
4. Water fill height to be recorded.
5. Out of the two adjacent pontoons identified for the puncture test, one has to be punctured and time has to be recorded along with the height of the water column coming out of the hole (this shall help in evaluating the final height of water after fill up of the punctured pontoon).
6. After water settles down in the first pontoon and no more significant change is seen, record the time taken and check for any leakage in the adjacent pontoon.
7. Repeat Steps 1 to 4 and puncture the next (adjacent) pontoon.
8. Repeat recording of the readings as in Steps 5 and 6.

9. After that, puncture the central deck and repeat the procedure and Steps 1 to 4.
10. Record the final readings of tilt and cross-check with theoretical values.

Procedure for Rainwater Fill Test

Blind the NRV and fill the central deck with water till 254 mm (10″) and keep the same for one day without and with the puncture test condition. The NRV shall be opened thereafter to see water flow from the roof drain and the time taken to remove the water.

Tests for Primary Drain

Drain pipes in a floating roof tank shall be pressure tested with water at 4 kg/cm^2g. Furthermore, during the floatation test, the roof drain valve shall be kept open and observed for leakage of the tank contents through the pipe drain.

Acceptance Criteria

For the pontoon puncture test, at each stage of water filling, the time taken in water settling and tilting shall conform to the approved floating roof design within mentioned tolerances of all the variables such as rim space, tilting, time taken to fill the pontoon, and so on.

For the rainwater fill test, the quantity of water to be filled on the roof shall be equivalent to 254 mm (10″) of rainfall over the entire tank area for a period of 24 hours. This shall be done with the primary drains closed, and when done so, water shall in no case come out of the outer rim of the deck. Immersion of the deck in water in that condition shall be as per design in a stabilized state; however, submergence of the outer rim shall not exceed 65% of its height at any point.

The time taken by roof drain water to come out from the central roof drain sump should be in conformance to the design and code requirements (Appendix C, API 650).

Note that after completion of the test, punctures should be blinded and LPT tested for leakage when the roof is at the lowest position (after hydrostatic testing) if welded, else coupling should be seal tightened.

Cleaning

After the water is drained, the tank shall be thoroughly cleaned, free from dirt and foreign materials, and shall be dried by air.

Records

Records shall be prepared as per the formats provided in Chapter 15 of this book.

Safety

All safety precautions shall be taken during the test.

Annexure H: Calibration of Tanks

General

The calibration of a storage tank is the process by which the volume in the tank is established in relation to the liquid height (up to the maximum fill height). There are a few methods available to measure the diameters of different shell courses at the site.

The API Measurement Committee on Petroleum Measurement issued a *Manual of Petroleum Measurement Standards* (MPMS) containing all the present individual measurement standards. Tank calibration methods are detailed in the second chapter, as follows:

- API MPMS 2.2A: *Measurement and Calibration of Upright Cylindrical Tanks by the Manual Strapping Method*
- API MPMS 2.2B: *Calibration of Upright Cylindrical Tanks Using the Optical Reference Line Method (ORLM)*
- API MPMS 2.2C: *Calibration of Upright Cylindrical Tanks Using the Optical-Triangulation Method (OTM)*
- API MPMS 2.2D: *Calibration of Upright Cylindrical Tanks Using the Internal Electro-Optical Distance Ranging Method (EODR)*
- API Standard 2555: *Method for Liquid Calibration of Tanks*

Calibration Requirements

It is preferred that all new tanks undergo calibration. However, for fire water tanks and the like, the accuracy of calibration is not so critical, and hence a theoretical calibration based on drawings would be good enough for the purpose. Regarding hydrocarbon storage tanks, the accuracy of calibration plays a role as the product prices are soaring.

For carrying out effective and meaningful calibration, note the following requirements:

- The tank shall have access inside for accurate determination of deadwood.
- Inspector must be able to ascertain datum plate (reference plate at the bottom) flatness and level check.
- Calibration shall be taken up only after successful hydrostatic testing of the tank.

Similarly, all tanks in service shall undergo recalibration or recomputation periodically, as follows:

- Recalibration shall be carried out at a set frequency or after repair.
- Frequency can be set by the owner or based on local regulations.
- General informative guidelines shall be as per API MPMS Chapter 2.2A.
- Recomputation is required only under certain specific conditions.

Recalibration Frequency

The Informative Appendix in API Chapter 2.2A provides guidelines regarding recalibration frequency. Recalibration is required on all tanks if internals are modified or the tank bottom repair work is undertaken. In addition, local regulations also play a vital role in deciding a recalibration frequency. On the basis of the 5/15 rule for tanks in custody service, the bottom shell course needs to be verified once every 5 years for diameter, thickness, and tilt if there are variations in diameter, thickness, and tilt from the previous calibration. If the computed volume based on new measurements is found in excess of 0.02% of the original calculated, recalibration is recommended. If it is found to be within 0.02%, verification of the bottom shell course every 5 years shall be continued until 15 years, after which total recalibration is recommended.

Working Tape Recalibration

The working tape used for calibration shall undergo recalibration after use on 20 tanks. Furthermore, it shall undergo recalibration if it is used on tanks whose circumference varies by more than 20% of the circumference of the tank on which the tape was originally calibrated. The master used for calibration shall be recertified once every two years.

Capacity Table and Raw Field Data

All raw data collated in the field shall be handed over to the tank owner along with the main capacity table generated based on data furnished by the owner and the field measurements taken by the contractor. The capacity table shall generally contain the following minimum information:

- Product ID, reference height, nominal diameter
- Product gravity, product temperature
- Critical zone
- Floating roof, total and incremental correction
- Shell temperature correction table if capacity table is prepared at 15°C or 60°F
- Appropriate footnote needs to be provided if corrections are already built into the table
- Reference height and reference gauge point location
- Method of calibration and date of calibration
- Certificate of calibration of working tape and master tape
- Signature of the certifying authority
- API Standard number (e.g., 2.2A) used for calibration

Recomputation

Recomputation or verification of the table is required when gravity changes by 10 deg API or higher. For this purpose, diameters from the last calibration report may be used to compute the new volumes resulting from gravity changes. Recomputation is required even when the average product temperature has changed by 20°F or more (if the temperature correction is built into the table).

Other General Information

It is not necessary for the tank to be empty for calibration. In fact it can be calibrated at any liquid level, but the tank shall not be calibrated during filling or draining. It is necessary for the tank level to be steady at the time of calibration, without any movement in and out. A tank can even be calibrated when the tank is full of water for a hydrostatic test. Upon completion of the hydrostatic test, the tank is calibrated with full of water, to be followed by destressing the tank (full emptying) to zero-stress condition and restressing the tank for the actual gravity of the product to be stored. For any given liquid level, the hydrostatic pressure is the greatest at the bottom and is a function of the specific gravity of the fluid stored. This results in marginal expansion of the tank. Therefore, this has to be accounted for. If it is not, it could impact the tank volume significantly depending on the diameter and thickness of the shell courses. Furthermore, for floating roof tanks, the compensation for buoyancy, which again is a function of gravity, also needs to be provided for. Yet another critical factor that affects the calibration is the product temperature, which results in expansion of the tank shell. The excess volume resulting from that could be significant.

Calibration is not generally recommended for tanks in insulated condition, especially for tanks meant for custody transfers and inventory.

Calibration Process Parameters

The following operational parameters shall be provided by tank owners to the calibration contractor and shall not be assumed under any circumstances by the calibration contractor:

1. Product temperature
2. Product-specific gravity
3. Roof leg position for floating roof tanks, as shown in the following sketch
4. Ambient temperature
5. Maximum fill height (depends on safety rules)

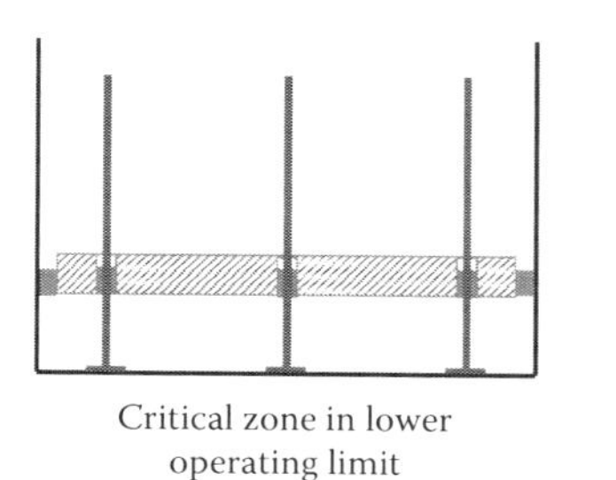

Critical zone in lower operating limit

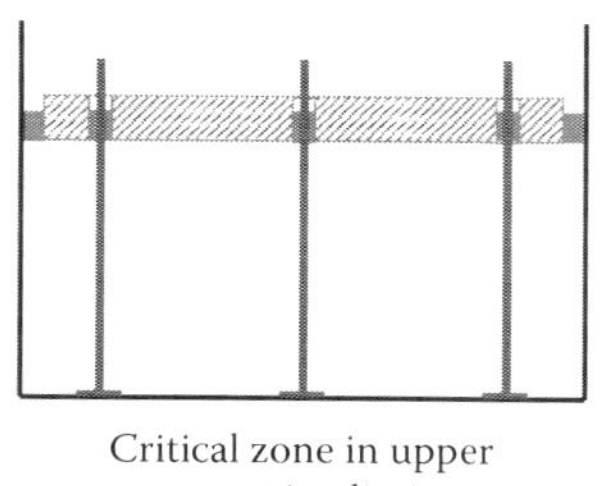

Critical zone in upper operating limit

Manual Method

Chapter 2.2A of API MPMS provides guidelines for the manual calibration method, and it also provides the basis for all other methods addressed by MPMS in other sections of Chapter 2.2. In the manual method, the following field measurements are involved.

Circumference Measurement of Each Shell Course

The circumference at each shell course shall be taken using working tape calibrated with appropriate tension. Multiple straps or a single strap may be used at each course based on the diameter. Tapes with a length of 30 m (100 ft) may be used for this measurement. The total number of straps thus required can be computed from $\pi D/30$ or $\pi D/100$ in meters or feet, respectively.

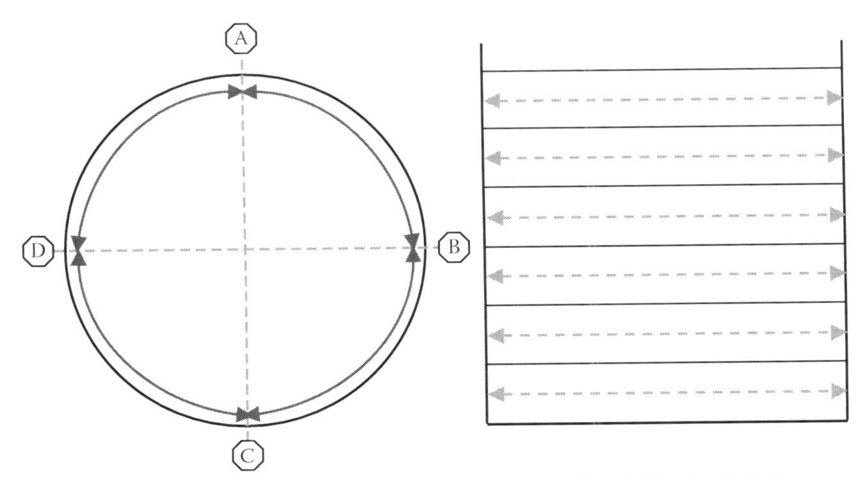

Four Straps used AB, BC, CD, DA, in each shell course at the middle of each shell course

Plate Thickness

The plate thickness of various shell courses can be measured ultrasonically all around each course (about 8 to 12 locations in each shell) and averaged for each course.

Diameter

The diameter needs to be computed based on the measured circumference and the thickness as specified previously.

Reference Height and Reference Gauge Point

This is a very critical component in tank calibration. For new tanks, this can be established easily, whereas for old tanks in service, access to the datum plate may not be possible, as the bottom shall usually be filled with solid sludge or other foreign materials. In case access is not available to measure the reference height again, it is better to use the reference height from the previous calibration table. The gauge point is the point from which gauging shall be undertaken, and hence this shall be clearly marked on the stilling well.

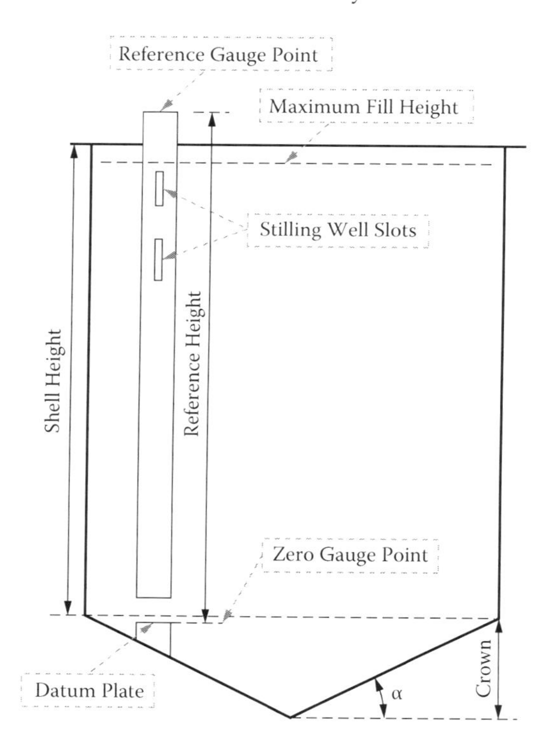

Critical Zone

In empty tanks, the roof leg position can be verified physically, whereas in the case of tanks in service, if access is not possible, this information may be taken from the last tank calibration table. Typically this could be in the range of 150 mm to 300 mm (6″ to 12″), but it is possible to be as high as 450 mm (18″) depending on the design of the floating roof.

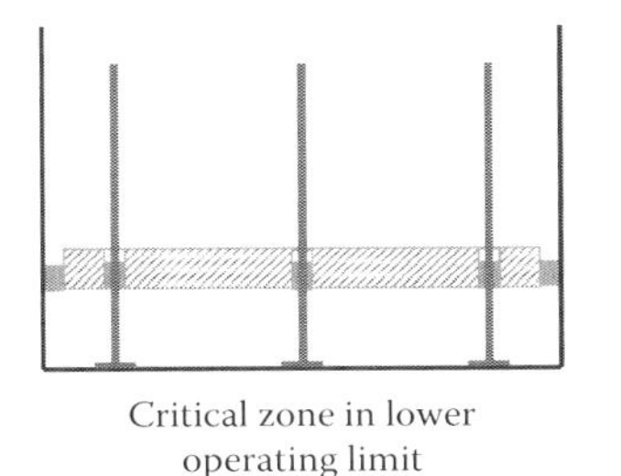

Critical zone in lower operating limit

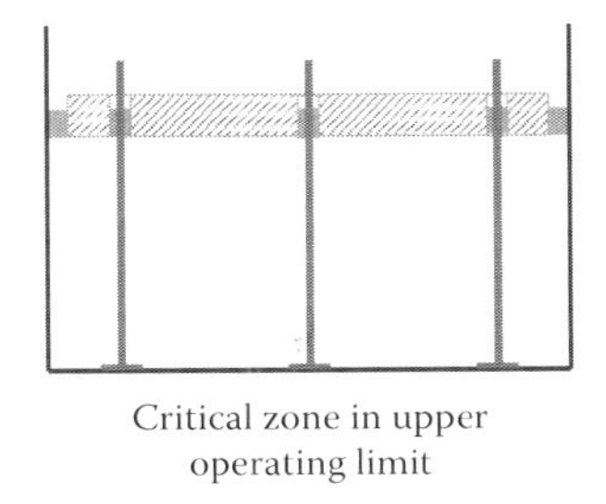

Critical zone in upper operating limit

Deadwood

The dimensions of all internal piping and other structures inside are physically measured, and their volumes are calculated distributed vertically from the datum plate. It is necessary to subtract the volume of deadwood as the tank calibration table is developed (volume vs. height). It shall be kept in mind that this is possible only when entry is permitted into the tank; if not, it shall be taken from the most recent calibration data available.

Floating Roof Weight

During calibration, the floating roof weight needs to be computed. This is collected either through physical measurement or from old tank table data. However, this calculation is prone to have large uncertainty on account of the quantum of weld that has been deployed on the roof and also the weight of attachments. Therefore, the best, most reliable, and most accurate source of information in this regard shall be based on the fabrication details of the floating roof tank to be collected from the tank manufacturer. Therefore, it is absolutely essential that these data be obtained from the manufacturer and retained by the owner of the tank for the entire service life of the tank. In addition, the history and details of repair carried out on the roof during the service life of the tank also have to be retained by the owner, since this also can affect the dead weight of the floating roof.

Maximum Shell Height

The maximum shell height is measured and documented as part of the development of the tank calibration table. This measurement is carried out on the external surface of the tank from the base.

Maximum Fill Height

The maximum fill height depends on the local conditions at the site where the tank is located. In earthquake-prone sites, the maximum fill height is restricted to 1,200 mm to 1,800 mm (4 ft to 6 ft) below the top rim or overflow. For floating roof tanks, the limiting factor is the maximum height to which the floating roof can go.

Bottom Calibration

Various types of tank bottoms are possible, meaning that they can be flat, cone up, or cone down. Tank bottoms need to be measured by a physical survey after entry is allowed for tanks in service. An alternate option is to calibrate the bottom with liquid (water). While the tank is in service, the zero gauge volume is copied from old tank calibration tables. The zero gauge volume is the volume of the tank below the datum plate.

Tilt

The tilt of a storage tank can be measured either using optical methods or manually by using plumb measurements.

Capacity Table

The capacity table for any storage tank is a table that gives the volume of the tank at any given height. The following corrections need to be applied to develop this table:

- Floating roof buoyancy correction
- Tank tilt correction
- Hydrostatic correction
- Shell temperature correction
- Master tape correction
- Working tape correction
- Other corrections such as tape rise

Floating Roof Buoyancy Correction

This correction is based on the gravity of the product and floating roof weight. Floating roof buoyancy (FRB) correction (in units for volume) shall be subtracted from the total volume at any given level as long as the floating roof is fully floating. In the critical zone, the FRB correction is distributed over the range of the zone, whereas below the critical zone, FRB correction is zero. The tank table carries the base FRB correction for a given gravity and incremental correction for variations in base gravity.

Tilt Correction

No correction is required when the tilt is less than 1 in 70, whereas tilt correction is required for tilts exceeding this value. However, the maximum tilt permitted is less than 2.4 in 100 (approximately 1 in 40).

Hydrostatic Head Correction

Hydrostatic head (liquid pressure) causes expansion of the tank shell. This is most severe at the lowest shell course. Additional volume resulting from pressure expansion of shell courses may be as high as 0.08%, depending on plate thickness. This volumetric expansion is a function of plate thickness and gravity for a given tank. API Chapter 2.2A provides a detailed procedure for calculating the incremental volume and the total volume for pressure correction. The additional incremental volume and total volume are generally included in the capacity table for a given gravity of the product. While the impact of variation in gravity up to +/− 5 deg API in computed volume is considered negligible, correction needs to be recomputed if the gravity change is more than 10 deg API.

Shell Temperature Expansion Correction

Because of the combined effect of ambient and product temperatures, the volume of the tank changes. The impact of this effect can be 0.05% and higher. The shell temperature determination equation provided in API Standard 2550 has been modified and is no longer the mean of the ambient and product temperatures. In the new equation, product temperature dominates. For insulated storage tanks, the shell and product temperatures are considered the same.

The temperature expansion factor may be included in the main capacity table for a given product and ambient condition, or the capacity table may be established at 15°C (60°F) and the shell temperature expansion factor may be applied externally for each batch received or discharged from the tank with actual field temperatures. Alternately, the capacity table may also be accompanied by a temperature expansion factor table when the capacity table is established at 15°C (60°F).

New Shell Temperature Equation

$Ts = (7*T_L+T_A)/8$ where T_L is liquid temperature and T_A is ambient temperature.

Master Tape Correction

Usually the tape calibration is carried out at 20°C (68°F), whereas the measured lengths shall be corrected to 15°C (60°F).

Other Corrections

Tape rise correction, if needed, also shall be applied.

Optical Reference Line Method

The reference standard for the optical reference line method (ORLM) used in the industry is API Chapter 2.2B, which provides guidelines for establishing the diameter of tanks by an optical method. Since an optical method is used. It can be applied internally or externally; external is easier.

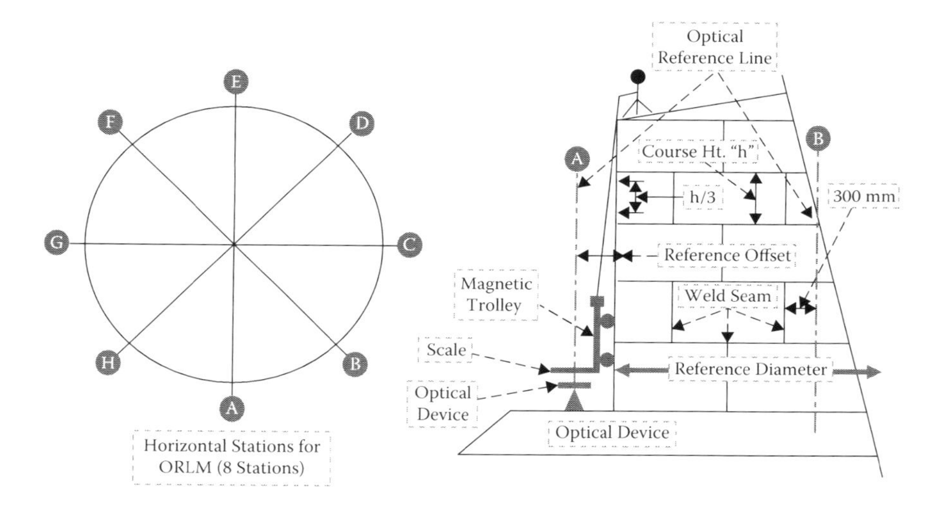

The method basically divides the tank into a number of horizontal and vertical stations. The number of horizontal stations may vary from 8 to 36 depending on the diameter of the tank. To measure the offset of the shell with reference to a reference level, a magnetic trolley with a graduated scale is moved vertically. The reference circumference of the bottom shell course is measured by the manual method (see API Chapter 2.2A). This is followed by the reference offset measurement made optically at the same height, where the reference circumference is measured. By moving the trolley upward in each of the designated stations (horizontally, A, B, C …), the course offset is measured optically at each horizontal station designated. The number of horizontal stations required is two per course. Later, the deviations in course offsets from the reference offset are averaged for each course. By using the reference circumference and averaged offset readings, the diameter of each shell course is established.

To ensure the reliability of measurements, the following aspects need to be given due consideration:

- The stability of the optical device shall be ensured.
- The device shall be level in all directions.
- The optical ray shall be vertical throughout the height of the tank (within limits).
- The reference offset shall be rechecked after the full vertical traverse of the magnetic trolley.
- The optical device shall be randomly checked for perpendicularity at three locations by rotating the device 360°.
- In extreme wind conditions, where it is difficult to maintain the trolley in contact with the shell, calibration shall not be undertaken

The rest of the measurements are identical to those specified in the manual method described in API Chapter 2.2A, and the tank capacity table shall be prepared based on these measurements.

The significant advantage of this process is that it does not require any scaffolding to carry out shell offsets. Furthermore, as the reference circumference is at the base (first shell course), it is much easier to take a proper accurate measurement.

Optical Triangulation Method

The reference standard followed in the industry for this optical triangulation method (OTM) is API Chapter 2.2C, which establishes diameters of the courses by an optical method. While this method can be applied internally and externally, it is easier to apply internally.

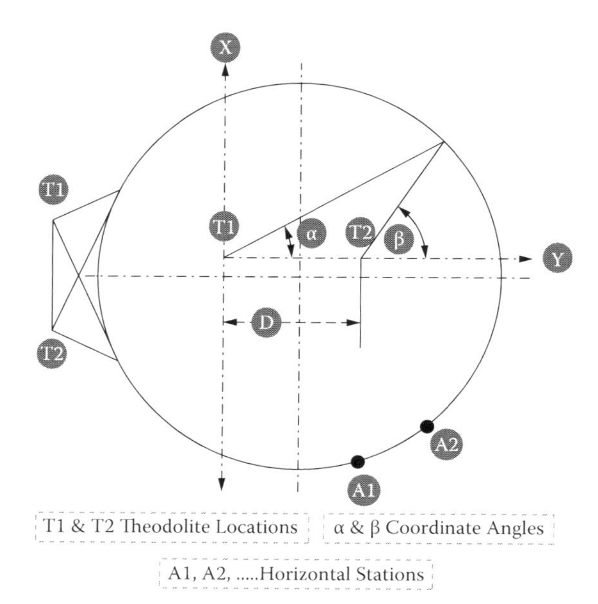

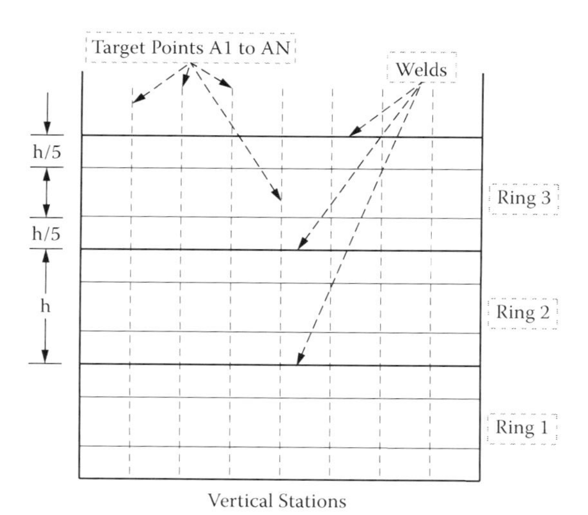

Here also, the tank is divided into horizontal and vertical stations for both internal and external methods. The tank profile shall be established by a triangle at each target point, as shown in the sketch above, and hence called the optical triangulation method (OTM). For the internal method, the reference distance D is established optically using temperature-compensated stadia typically 2 m long. Subsequently tank coordinates A(x, y) are measured optically using two theodolites (T_1 and T_2). For the external method, the tangential angles are measured along with the distance between the two theodolites (T_1 and T_2). Subsequently, the diameters are computed using mathematical computational procedures.

The following aspects for carrying out measurements by OTM shall be given due consideration:

- The stability of optical devices shall be ensured.
- Devices shall be level in all directions.
- Distance D for the internal method shall be measured again at the end.

As in the case of ORLM, in this method the rest of the measurements are identical to those specified in the manual method described in API Chapter 2.2A, and the tank capacity table shall be prepared based on these measurements.

Electro-Optical Distance Ranging Method

The reference standard for the electro-optical distance ranging method (EODR) used in the industry is API Chapter 2.2D. While ORLM and OTM can be applied both internally and externally, the EODR method can be applied only internally. Like ORLM and OTM, this method also establishes the diameters of all shell courses.

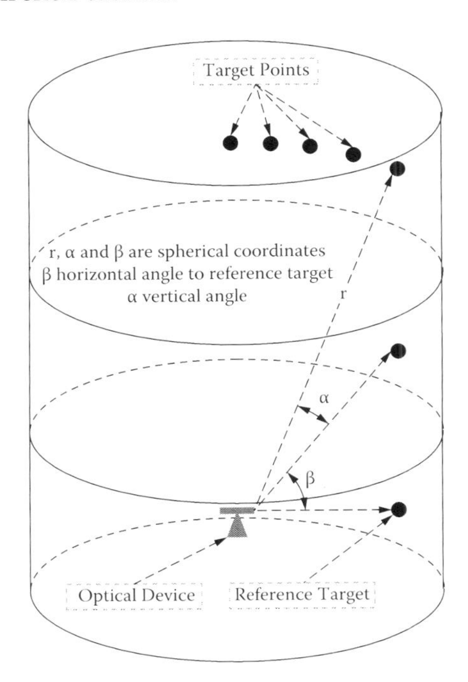

In this method, a reference target on the bottom course is established followed by observing the reference distance and reference angle. Spherical coordinates are measured using a distance ranging device (r, α, β) for each target point. In this method, the tank profile is thus established from bottom to top. The reference distance of the target and the reference angle of the target at the end are rechecked. Based on these measurements, using standard mathematical procedures, the diameter of individual shell courses is computed. With an online computer, diameters can be determined instantaneously.

The following aspects for carrying out measurements with the EODR method shall be given due consideration:

- The device shall be level in all directions.
- Measurements at the reference targets after the complete traverse of the tank shall be rechecked, and measurements thus obtained shall be a repetition of the initial measurements.

As in the case of the other three methods, in this method also the rest of the measurements are identical to those specified in the manual method described in API Chapter 2.2A, and the tank capacity table shall be prepared based on these measurements.

Liquid Calibration

The applicable reference standard for this method is API Standard 2555. Many client specifications for hydrocarbon storage tanks require liquid calibration. This is a very accurate method, but it is time-consuming. In this method, level versus volume is established directly. Volume Q1 is metered (volume is measured through a flow meter, calibrated prior to the start of tank calibration), and the corresponding level L1 is measured. Further increments in levels shall depend on the tank diameter and generally should be 10 mm to 300 mm (6″ to 1′). The flow meter used for tank calibration shall undergo calibration before and after calibration of the tank to ensure accuracy.

In liquid calibration, hydrostatic correction is not necessary, because at each level, the tank gets expanded because of the internal pressure. Furthermore, this procedure does not require any deadwood correction. For liquid calibration, either the product or water can be used. If water is used for calibration, adjustments to the volume by courses are necessary because of variation in gravity between water and the product. As in the case of other procedures, the reference height shall be measured per API Chapter 2.2A.

Summary

It is required that tanks used for custody transfers, mass balance in refineries, volume balances in tank farms and pipeline terminals, and so on are calibrated. In this regard the following aspects shall be given due consideration irrespective of the method adopted:

- Tank fabrication drawings shall not be used as a basis for determining the tank diameter.
- Recalibration at a set frequency is also equally important.
- Any of the previously mentioned methods may be used to establish tank diameters.
- Tank calibration shall never be undertaken over insulation in insulated tanks.
- For insulated tanks, internal calibration or liquid calibration may be used if insulation cannot be removed.
- If insulation can be removed, external calibration may be used.
- Shell expansion due to hydrostatic pressure and expansion due to temperature are not negligible and hence shall be included in the development of the tank capacity table.

Annexure I: Floating Roof Drains

API 650 (Annex C) permits three types of primary roof drains as follows:

1. Manufacturer's standard drain
2. Steel swing or pivot-jointed pipe drains, designed and packed for external pressure
3. Stainless steel armored hose

Most of the clients in the oil and gas industry prefer options 2 and 3, as they have a proven track record.

Articulated Drain System

Articulated drain systems are composed of steel pipes with swivel joints designed to withstand the forces they may be subjected to under all operating regimes. The system usually incorporates heavy-duty swivel joints with a lot of flexibility to accommodate client preferences. (Pictures courtesy M/s Ateco Tank Technologies, Turkey.)

Swivel joint articulated drain system.

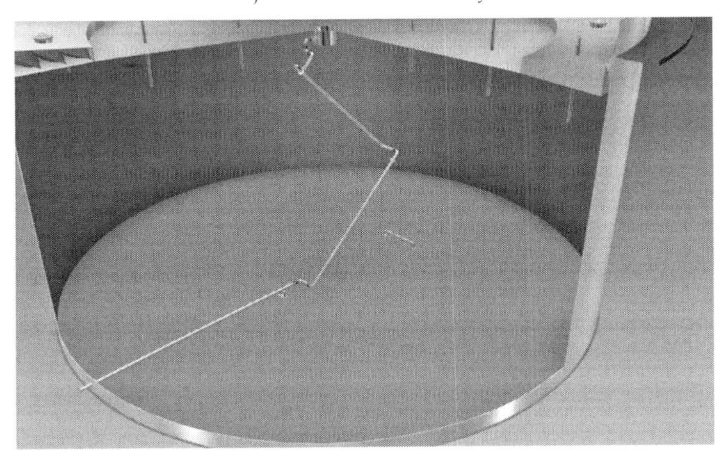

Swivel joint articulated drain system.

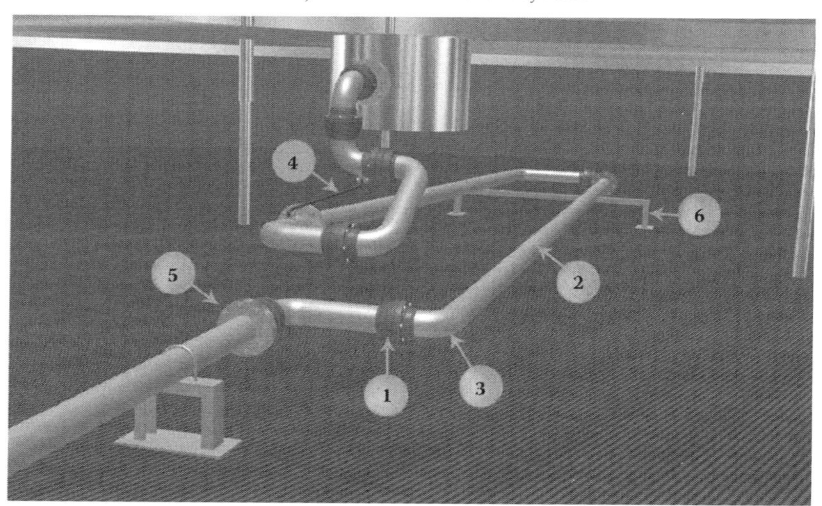

Part No.	Description	Material
1	Swivel joint	Carbon steel (CS) or stainless steel (SS) body
2	Pipe work	CS AST A 106 Gr B Sch 40 smls
3	Elbows	CS ASTM A 234 WCB Sch 40 smls
4	Link chain	SS
5	Flanges	CS ASTM A 105 #150 WNRF
6	Base leg supports	CS

System type	Articulated roof drain system
Description	Steel pipe system with rotary swivel joints
Used on	External floating roof tanks
Service	Suitable for a wide range of products
API 650	Compliant
Advantages	Robust construction, no maintenance, no issues related to contacting roof legs
Disadvantages	High initial cost, longer installation time

Flexible Hose Drain System

Extremely flexible and easy to handle or bend hoses are used as in the following sketches to drain water from a floating roof. Drain hoses are specially designed to resist immersion in corrosive media with adequate flexibility to take care of all possible movements of a floating deck. The hose is specifically designed to have a negative buoyancy to prevent floatation of the hose when empty. Usually hoses are externally swaged with stainless steel ferrules and Viton seals.

Flexible hose drain system.

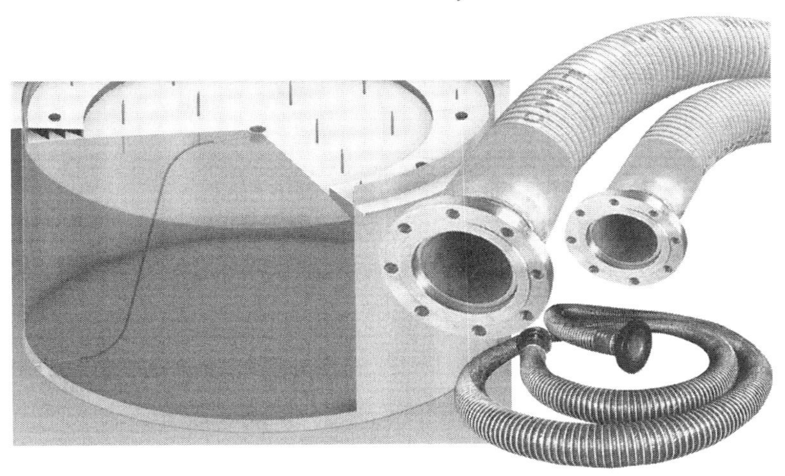

Flexible hose drain system.

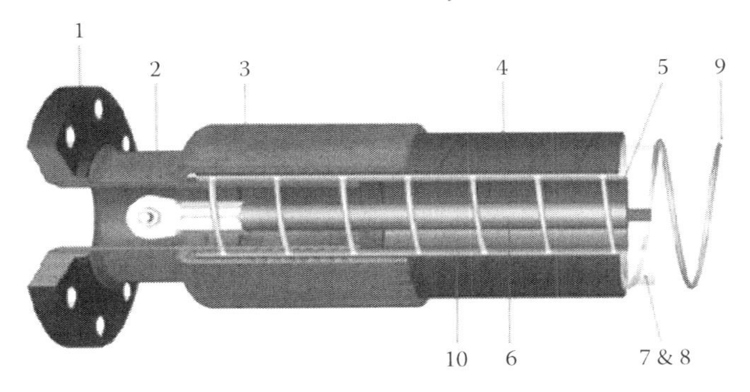

Part No.	Description
1	ANSI #150 flange
2	Hose nipple
3	Hose ferrule
4	Hose outer cover
5	Hose inner cover
6	Lead ballast (with steel inner core)
7	Inner reinforcing ply
8	Outer reinforcing ply
9	High tensile reinforcing wire
10	White reinforcing stripe

Annexure J: Floating Roof Seal Selection Guidelines

While numerous solutions are available in the market with regard to the type and number of seals to be provided on floating decks, only very limited guidance is available from standards specifications. If the wrong types of seals are selected, issues with regard to emissions and outages due to premature failures can be expected. Therefore, it was considered essential to provide some more insights about seal selection criteria and also about commercially available seals and their working principles, from which the readers might be able to decide on the types of seals they may have to select for their specific situation.

Considerations

The seal selection process shall definitely consider the following:

- Emission mitigation
- Centering of floating deck and smooth up-and-down movement of floating deck
- Safeguards against probable rim fires
- Ease of maintenance and inspection

Chemical Compatibility

All metallic and nonmetallic materials (including elastomers and fabric) used in the seal shall be compatible with the fluid stored. Apart from the currently proposed fluid for storage, consideration of probable changes in the quality of products that may arise in the future also might be rewarding in the long run.

UV and Weather Resistance

Ultraviolet (UV) and weather resistance is another key consideration to seal selection and design. Materials exposed to weather should be resistant to the effects of sun, wind, and, depending on where the tank is located, extreme cold and/or heat, especially the dust storm conditions prevailing in the Middle East.

Resistance to Abrasion

The seal's resistance to abrasion is yet another consideration, which depends on factors such as cycle frequency, service life, and condition of abrading surface (shell surface). In addition, seal construction features such as the width of the plate, pressure application method, and so on can also significantly affect the performance and service life of the seal.

Suitability against Operational Range

Since the tank shell is a fabricated structure using comparatively low thickness plates and because of the huge quantities of welding involved, it is not possible to obtain precise accuracy with regard to dimensions. Therefore, the suitability of the roof seal to accommodate a wide range of dimensional disparity would be an added feature of any seal. In addition, for external floating roofs where dynamic forces such as wind and turbulence can move the floating roof, a higher range compared to that for internal floating roofs may be called for.

Similarly, designing seals to operate effectively across varying rim spaces is extremely important. When rim space varies greatly, support arms in hinged shoe seals should be long enough relative to the rim space to minimize shoe drop. In addition, secondary seal plate length, gauge, and tip type should be determined only after careful evaluation of the operating conditions including cycle frequency, temperature, and shell condition.

Flexibility along Circumference

Circumferential flexibility is the flexibility of the seal circumferentially and is closely related to flexibility for rim space variations. Seal designs and construction should allow expansion and contraction without compromising seal integrity. Using modular vapor barriers such as gaskets can create gaps. Seals should use a continuous vapor barrier to ensure a proper seal.

Accommodation of Shell Irregularities

A good seal shall be flexible and apply continuous pressure even when the shell surface has irregularities, such as ovality and local departure from profile. To counter such issues, seal vendors recommend a flexible shoe with a pressure system that applies pressure consistently in multiple points across the shoe, distributed both horizontally and vertically. For wiper-type seals, the tip shall be flexible and durable, and continuous pressure shall be applied to prevent gapping. Shell surface variations can present quite a challenge to maintaining a tight seal.

Centering of Floating Deck

A floating roof seal shall apply enough pressure to keep the floating roof centered. If this is not centered, it can drift in external wind or turbulence or drag forces in product flow, which can result in catastrophic failure of the roof. To keep floating roofs centered, the pressure mechanisms in floating roof seals shall be substantial and made from materials that do not yield or degrade over time.

Easiness in Maintenance

Seals shall be easy to clean and made gas free. Traditional foam log, bag, and tube seals have the potential to trap hydrocarbons, which creates an unsafe environment for maintenance workers and also poses environmental issues with regard to disposal.

Additional considerations include seal fabrics and wax scrapers. Seal fabrics shall be chosen to ensure chemical compatibility and durability under specific environmental and operating conditions and also shall be fire retardant. Wax scrapers may be a consideration if the product creates waxy buildup on the tank shell. If this buildup is allowed to occur, the secondary seal can partially scrape some wax off when the deck travels upward, resulting in hydrocarbons on the roof. This product accumulation not only is a fire hazard and large source of emissions but can also lead to plugged foam dam weep holes and topside corrosion of the floating roof. The effective solution would be the placement of wax scrapers below the liquid level. These scrapers shall be made of hardened stainless steel (to prevent yielding) and be designed with additional pressure application located just above the wax scraper.

Good seal selection and design start with good data and careful planning. Accurate tank data including rim space and verticality surveys would greatly benefit the seal vendor to ensure that the seal system is designed to effectively negotiate the rim space, shell irregularities, cycle frequency, operating conditions, and any additional local regulatory requirements. Therefore, a good liaison with the seal vendor in all probability could provide right and long-lasting solutions.

Types of Seals

(Pictures and details courtesy M/s Ateco Tank Technologies, Turkey.)

1. Primary scissor-type mechanical shoe seal.

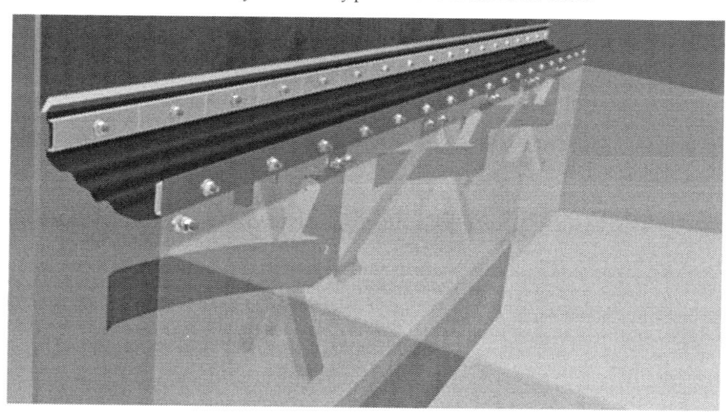

BASIC INFORMATION	
TYPE	Primary seal
DESCRIPTION	Mechanical scissor shoe seal
USED ON	External floating roof tanks
RIM SPACE	Nominal rim space of 200 mm ± 125 mm or 300 mm ± 225 mm
SERVICE	Suitable for all products with correct material selection
API COMPLIANT	Yes
CODRES COMPLIANT	Yes
API 2003 COMPLIANT	Yes
AROMATIC SERVICE	100%
TYPICAL SERVICE LIFE	15–25 years
WEIGHT	23 kgs/mt (typically based on 200 mm rim space)

2. Primary pantograph-type mechanical shoe seal.

BASIC INFORMATION	
TYPE	Primary seal
DESCRIPTION	Pantograph-type mechanical shoe seal
USED ON	External floating roof tanks
RIM SPACE	Nominal rim space of 200 mm ± 125 mm or 300 mm ± 225 mm
SERVICE	Suitable for all products with correct material selection
API COMPLIANT	Yes
CODRES COMPLIANT	Yes
API 2003 COMPLIANT	Yes
AROMATIC SERVICE	100%
TYPICAL SERVICE LIFE	25–40 years
WEIGHT	43 kgs/mt (typically based on 200 mm rim space)

3. Low-profile secondary seal LP series.

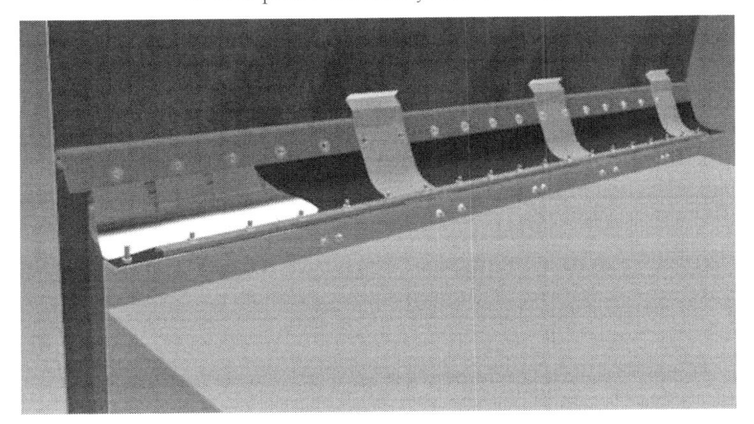

BASIC INFORMATION	
TYPE	Secondary seal
DESCRIPTION	Low-profile secondary seal LP series
USED ON	External floating roof tanks
RIM SPACE	Nominal rim space of 200 mm ± 125 mm or 300 mm ± 225 mm
SERVICE	Suitable for all products with correct material selection
API COMPLIANT	Yes
CODRES COMPLIANT	Yes
API 2003 COMPLIANT	Yes
AROMATIC SERVICE	100%
TYPICAL SERVICE LIFE	15–25 years
WEIGHT	7.5 kgs/mt (typically based on 200 mm rim space)

4. Secondary seal: flat wiper lip.

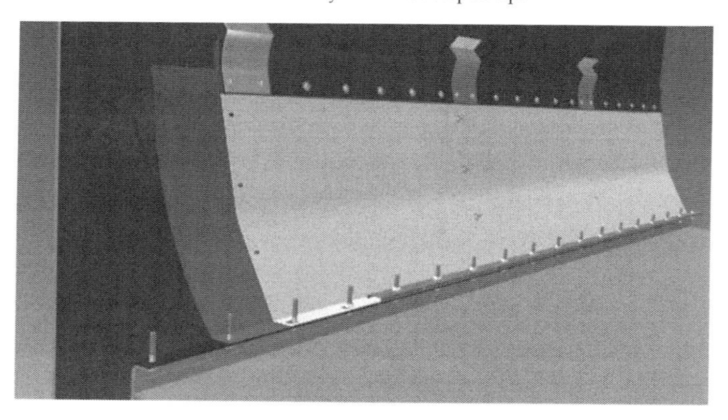

BASIC INFORMATION	
TYPE	Secondary seal
DESCRIPTION	Rim-mounted compression-plate-type secondary seal
USED ON	External floating roof tanks
RIM SPACE	Up to 250 mm nominal. Values in excess of this require modifications to the pontoon rim.
SERVICE	Suitable for all products with correct material selection
API COMPLIANT	Yes
CODRES COMPLIANT	Yes
API 2003 COMPLIANT	Yes
AROMATIC SERVICE	100%
TYPICAL SERVICE LIFE	15–25 years
WEIGHT	11.5 kgs/mt (typically based on 200 mm rim space)

5. Secondary seal: L wiper lip.

BASIC INFORMATION	
TYPE	Secondary seal
DESCRIPTION	Rim-mounted compression-plate-type secondary seal
USED ON	External floating roof tanks
RIM SPACE	Up to 250 mm nominal. Values in excess of this require modifications to the pontoon rim.
SERVICE	Suitable for all products with correct material selection
API COMPLIANT	Yes
CODRES COMPLIANT	Yes
API 2003 COMPLIANT	Yes
AROMATIC SERVICE	100%
TYPICAL SERVICE LIFE	15–25 years
WEIGHT	12.5 kgs/mt (typically based on 200 mm rim space)

6. Secondary seal: dual wiper lip.

BASIC INFORMATION	
TYPE	Secondary seal
DESCRIPTION	Rim-mounted compression-plate-type secondary seal
USED ON	External floating roof tanks
RIM SPACE	Up to 250 mm nominal. Values in excess of this require modifications to the pontoon rim.
SERVICE	Suitable for all products with correct material selection
API COMPLIANT	Yes
CODRES COMPLIANT	Yes
API 2003 COMPLIANT	Yes
AROMATIC SERVICE	100%
TYPICAL SERVICE LIFE	15–25 years
WEIGHT	10.5 kgs/mt (typically based on 200 mm rim space)

7. Double seal: flat wiper lip.

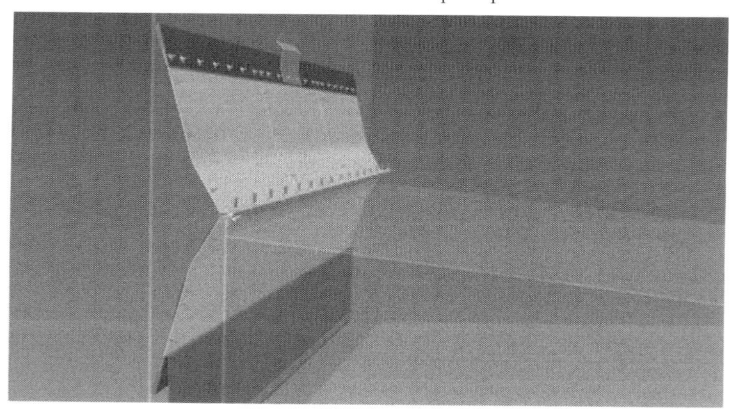

BASIC INFORMATION	
TYPE	Double seal
DESCRIPTION	Rim-mounted compression-plate-type double seal
USED ON	External floating roof tanks
RIM SPACE	Up to 250 mm nominal. Values in excess of this require modifications to the pontoon rim.
SERVICE	Suitable for all products with correct material selection
API COMPLIANT	Yes
CODRES COMPLIANT	Yes
API 2003 COMPLIANT	Yes
AROMATIC SERVICE	100%
TYPICAL SERVICE LIFE	15–25 years
WEIGHT	25 kgs/mt (typically based on 200 mm rim space)

8. Double seal: L wiper lip (1).

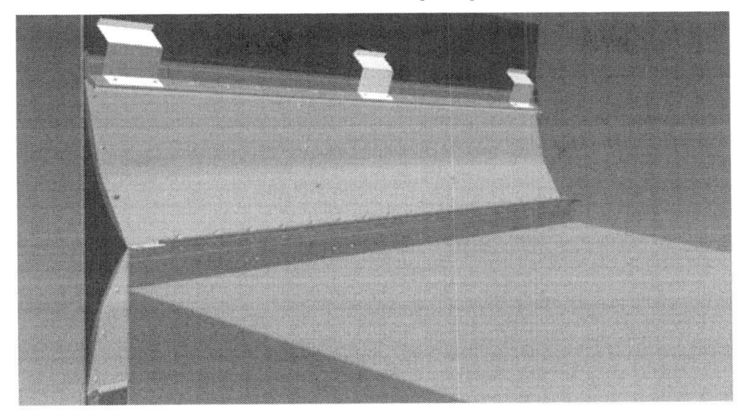

BASIC INFORMATION	
TYPE	Double seal
DESCRIPTION	Rim-mounted compression-plate-type double seal
USED ON	External floating roof tanks
RIM SPACE	Up to 250 mm nominal. Values in excess of this require modifications to the pontoon rim.
SERVICE	Suitable for all products with correct material selection
API COMPLIANT	Yes
CODRES COMPLIANT	Yes
API 2003 COMPLIANT	Yes
AROMATIC SERVICE	100%
TYPICAL SERVICE LIFE	15–25 years
WEIGHT	25 kgs/mt (typically based on 200 mm rim space)

9. Double seal: L wiper lip (2).

BASIC INFORMATION	
TYPE	Double seal
DESCRIPTION	Rim-mounted compression-plate-type double seal
USED ON	External floating roof tanks
RIM SPACE	Up to 250 mm nominal. Values in excess of this require modifications to the pontoon rim.
SERVICE	Suitable for all products with correct material selection
API COMPLIANT	Yes
CODRES COMPLIANT	Yes
API 2003 COMPLIANT	Yes
AROMATIC SERVICE	100%
TYPICAL SERVICE LIFE	15–25 years
WEIGHT	25 kgs/mt (typically based on 200 mm rim space)

10. Foam graver seal.

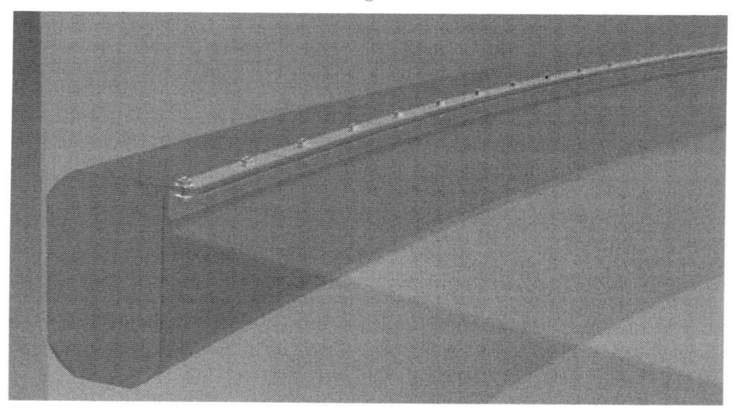

BASIC INFORMATION	
TYPE	Primary seal
DESCRIPTION	Vapor or product-mounted-type foam-filled primary seal
USED ON	External floating roof tanks or internal pan deck floating roof tanks
RIM SPACE	Nominal rim spaces of 120 mm, 150 mm, 200 mm, and 250 mm
SERVICE	Suitable for all products with correct material selection
API COMPLIANT	Yes
CODRES COMPLIANT	Yes
API 2003 COMPLIANT	Yes
AROMATIC SERVICE	100%
TYPICAL SERVICE LIFE	10–15 years
WEIGHT	7 kgs/mt (typically based on 200 mm rim space)

11. Foam deltoid seal.

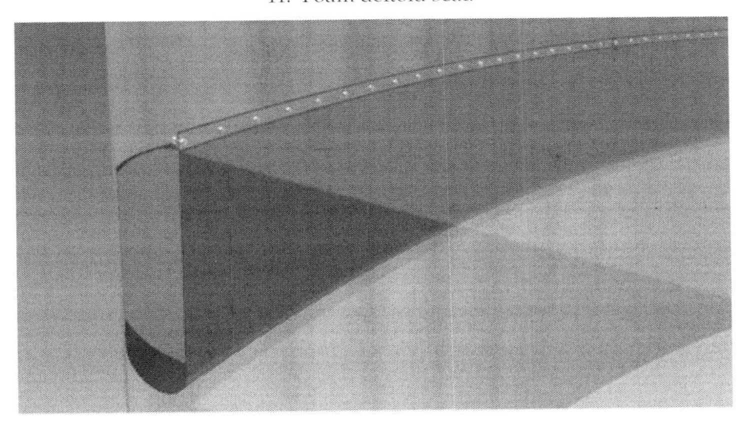

BASIC INFORMATION	
TYPE	Primary seal
DESCRIPTION	Product-mounted-type foam-filled primary seal
USED ON	External floating roof tanks or internal pan deck floating roof tanks
RIM SPACE	Nominal rim spaces of 125 mm, 150 mm, 200 mm, and 270 mm
SERVICE	Suitable for all products with correct material selection
API COMPLIANT	Yes
CODRES COMPLIANT	Yes
API 2003 COMPLIANT	Yes
AROMATIC SERVICE	100%
TYPICAL SERVICE LIFE	10–15 years
WEIGHT	12 kgs/mt (typically based on 200 mm rim space)

12. Integral foam dam.

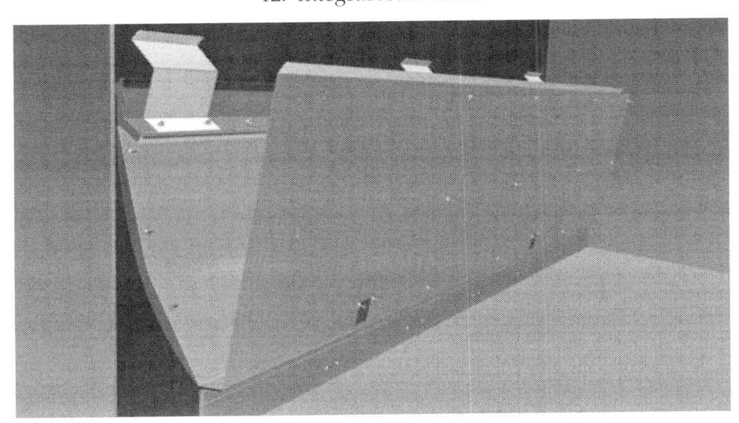

BASIC INFORMATION	
TYPE	Foam dam
DESCRIPTION	Rim-mounted integral-type foam dam
USED ON	External floating roof tanks
RIM SPACE	N/A
SERVICE	N/A
API COMPLIANT	Yes
CODRES COMPLIANT	Yes
API 2003 COMPLIANT	—
AROMATIC SERVICE	—
TYPICAL SERVICE LIFE	20–25 years
WEIGHT	17 kgs/mt (typically based on 200 mm rim space)

13. Foam seal (1).

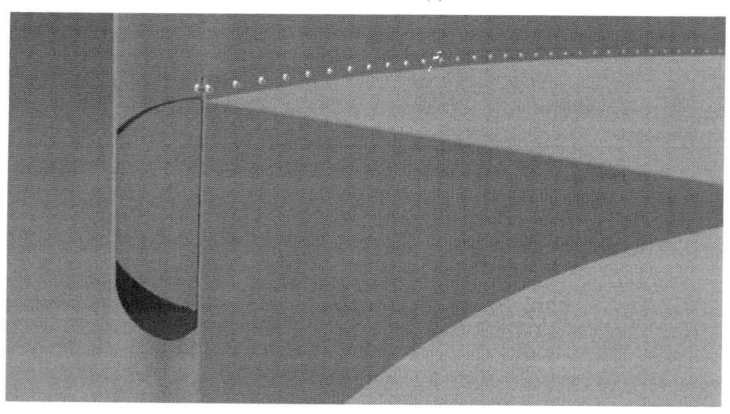

BASIC INFORMATION	
TYPE	Primary seal
DESCRIPTION	Vapor-mounted-type foam-filled primary seal
USED ON	External floating roof tanks
RIM SPACE	Nominal rim spaces of 200 mm
SERVICE	Suitable for all products with correct material selection
API COMPLIANT	Yes
CODRES COMPLIANT	Yes
API 2003 COMPLIANT	Yes
AROMATIC SERVICE	100%
TYPICAL SERVICE LIFE	10–15 years
WEIGHT	11 kgs/mt (typically based on 200 mm rim space)

14. Foam seal (2).

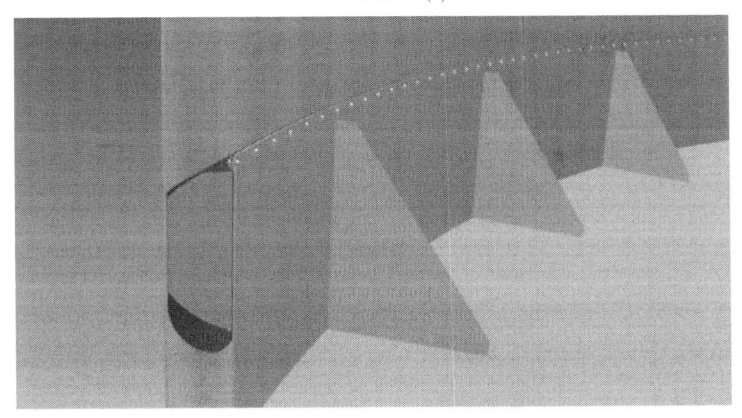

BASIC INFORMATION	
TYPE	Primary seal
DESCRIPTION	Vapor or product-mounted-type foam-filled primary seal
USED ON	External floating roof tanks or internal pan deck floating roof tanks
RIM SPACE	Nominal rim spaces of 115 mm, 140 mm
SERVICE	Suitable for all products with correct material selection
API COMPLIANT	Yes
CODRES COMPLIANT	Yes
API 2003 COMPLIANT	Yes
AROMATIC SERVICE	100%
TYPICAL SERVICE LIFE	10–15 years
WEIGHT	7 kgs/mt (typically based on 200 mm rim space)

15. Foam seal (3).

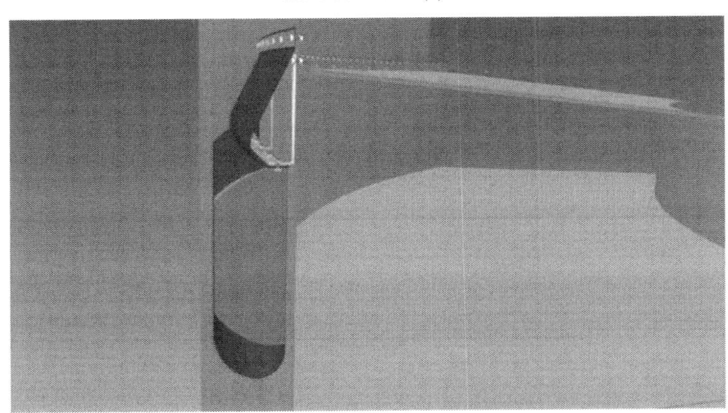

BASIC INFORMATION	
TYPE	Primary seal
DESCRIPTION	Product-mounted-type foam-filled primary seal
USED ON	External floating roof tanks
RIM SPACE	Nominal rim spaces of 200 mm
SERVICE	Suitable for all products with correct material selection
API COMPLIANT	Yes
CODRES COMPLIANT	Yes
API 2003 COMPLIANT	Yes
AROMATIC SERVICE	100%
TYPICAL SERVICE LIFE	10–15 years
WEIGHT	20 kgs/mt (typically based on 200 mm rim space)

16. Tube liquid-filled bag seal.

BASIC INFORMATION	
TYPE	Primary seal
DESCRIPTION	Product-mounted-type liquid-filled primary seal
USED ON	External floating roof tanks
RIM SPACE	Nominal rim space up to 275 mm
SERVICE	Suitable for all products with correct material selection
API COMPLIANT	Yes
CODRES COMPLIANT	Yes
API 2003 COMPLIANT	Yes
AROMATIC SERVICE	75% or 100%
TYPICAL SERVICE LIFE	10–15 years
WEIGHT	16 kgs/mt (typically based on 200 mm rim space) Weight includes lower mounting angle but excludes sealing liquid

Photographs

Pantograph-type mechanical shoe seal.

Secondary seal system.

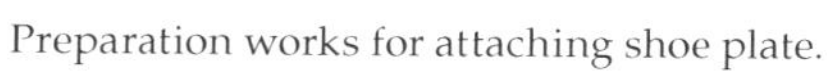
Preparation works for attaching shoe plate.

Installation works of shoe arm and counterweight arm.

Installed shoe arm and counterweight arm.

Shoe plate attachment works.

Shoe plates after installation.

Bibliography

1. *Aboveground Storage Tanks*, Philip E. Myers.
2. *Above Ground Storage Tank Design, Inspection, and Compliance Management*, a presentation by Gregory Coppola, P.E., Kinder Morgan Energy Partners.
3. *Pictorial Surface Preparation Standards for Painting Steel Surfaces*, HMG Paints.
4. *NFPA 30*
   Flammable and Combustible Liquids Code, 2003 Edition, National Fire Protection Association.
5. *Specifications for Storage Tanks*, KLM Technology Group.
6. Brochure of M/s Bygging Udheman.
7. Brochure of M/s Byggwik (UK) Ltd.
8. Brochure of Kansai Protection Coating Systems, Kansai Paints Japan.
9. Brochure of Ateco Floating Roof Seals.
10. *Tank Erection Procedure*, M/s Bygging (India) Ltd.
11. Brochure of Ateco Roof Drains.
12. Article on Surface Preparation, M/s Trancocean.
13. ASME Section VIII Div (1). Rules for construction of Pressure Vessels.
14. *Welded Steel Tanks for Oil Storage*, API 650.
15. *Design and Construction of Large, Welded, Low-Pressure Storage Tanks*, API 620.
16. *Tank Inspection, Repair, Alteration, and Reconstruction*, API 653.
17. *Measurement and Calibration of Upright Cylindrical Tanks by the Manual Strapping Method*, API MPMS 2.2A.
18. *Calibration of Upright Cylindrical Tanks Using the Optical Reference Line Method (ORLM)*, API MPMS 2.2 B.
19. *Calibration of Upright Cylindrical Tanks Using the Optical-Triangulation Method (OTM)*, API MPMS 2.2 C.
20. *Calibration of Upright Cylindrical Tanks Using the Internal Electro-Optical Distance Ranging Method (EODR)*, API MPMS 2.2 D.
21. *Method for Liquid Calibration of Tanks*, API Standard 2555.
22. *Fire Precautions at Petroleum Refineries and Bulk Storage Installation*, Institute of Petroleum.

Index

Q

R

S

T

X

Y

Z

Printed in the United States
by Baker & Taylor Publisher Services